My Journey Through
Nick Yarri

Copyright © 2017 Nick Yarris

All rights reserved.

ISBN:1976456436
ISBN-13:9781976456435

DEDICATION

His real name is Michael Yarris but I always call him Poppy. Born in May of the year 1934 when times were hard, he sure has seen a lot of life to reach the age of 83 now. 'Mike Yarris' to his friends, (and to those really old friends), he was also called 'Buck' because of how in all the many years of being a hunter, he never killed a male deer. My cousins and uncles or all of the men who we hunted along side of back in Pennsylvania affectionately called him this name at times. Me? I am just glad that he was not called this because he had killed so many deer. To my way of thinking, "Good on you Poppy, I am glad you never killed one male deer *ever*".

I began interviewing my dad on the sly earlier this year. I was making one more epic journey in life to go back to the USA with Laura and her 2 daughters Bethany and Zara. I wanted to keep my father posted on what I was up to as things have a way of going wildly for me. I like for him to know what is going down as best that I am able to without it being a worry. He has seen me go through every hurt since my release from prison without much judgment upon me thankfully. I know, and he knows that basically now that my mother Jayne Yarris passed away in 2011, how he too has little time left in life. I am glad it is not filled with he and I being at odds.

I concede that my relationship with him is a strange one. I know it as much as he does. We talk on the phone and I share with my father things that I am doing or where I am as I said, but it's never much more involved than that. I just want whatever time remaining to be good between us and he is tired of ever mentioning the past.

So partly out of curiosity and as a way to have something to talk about on his side, I began to interview him about his life. It is not easy as he won't tell you things if asked in terms of any parts being used in my writing. I already know that he won't like being mentioned again here for what it brings up inside of him to think about.

When I ring my father up on the phone I try to always get him to laugh and I tempt him into being facetious or humorous when I have the chance. He has terrible pain daily, so I have to try and sneak my questions in on him for my hidden efforts to get information out him to pay off without burdening him emotionally.

I wait for him to get past the broken down body complaints, on into and then through the laments of his not being able to breathe very well any more. Once we each curse the sports teams that we each love to hate, we then segue into a moment of calm. Then I ask things like: "Poppy, what was your first car"?

Then comes the story of his which I so cherish, and as he begins to speak while drifting into memory, I smile for knowing that I have gotten him past thoughts of missing my mother, taking him from the many horrors he can still recall of things which went badly for our family...

He is such a wonderful storyteller. He always has a charming sense of making the tale come alive in a sweetly bold way. His first car? Oh it was a 1951 Chevy! The way he tells it goes like this...

It was a big ol' Black sedan and it was fast and sexy he said. My father got this car when my Uncle Nicky, whom I am named for, went one day with my father after they had sold some pigs they owned and got some cash together.

You see, Mike Yarris was a first generation born Russian-American. My grandfather was a pig farmer who lived in an area since he came to America just after the Russian revolution. His farm was on land that is part of what is now the Philadelphia Airport.

My grandfather was to die early on in my fathers' life, just like how my grandmother who had bore him had died too. My grandmother actually died in a flood that came inside of her home from Darby creek that was overflowing at the time in Pennsylvania. All of it happened when my dad wasn't more than like 15 years old. The farm was left to my dad as well as to my uncles John and Nicky.

My Uncle John left shortly after all this when he got married and soon it was just the two youngest brothers left with the farm.

How crazy that at the age of just 16, and my uncle being only 18 years old as well, that they went over to a car dealership just outside of Philadelphia where they put down a deposit of 750.00 U.S. Dollars for a new car.

With just a promise to the dealership that he and my uncle would come back in a few months with the remaining 1.500.00 dollars that they expected from further pig sales, they drove off in a brand new car! Poppy added that back then, dealerships believed in people, and how there was no endless papers and finance people to call up. You took your word serious or you were ashamed to be in debt.

To be in debt was a shameful thing back then on two levels Poppy said to me. It meant mostly that you had no money and, or that you had no respect for those whom you owed. No, you did not seek credit unless you were poor or hurting and you needed it short term.

Nowadays he says that all of the rich folks use credit as a status symbol so that the poor ones cannot get any money loaned to them. In so many ways he is right. Nobody is ashamed to be in debt it seems.

Anyway, Poppy said that he had many girlfriends in that car and he really brightened up telling me innuendo style, how many girls loved to be in the back seat making out with him.

Just hearing his voice come alive momentarily made my heart swell with affection for him. So I ventured further then when I asked him to tell me how he had met my mother. Wow, I did not expect this one...

Poppy, at the ripe old age of something like 20 years old is quite the man's man by that point. The farm was lost to the city after my uncle and my father let their youth squander it all. So now my dad had a normal job as a roofer in the area.

One day he walked into a place on Elmwood avenue in Philadelphia called "Rosie's Tavern" that was located by Island road. There was my mother sitting there in a booth sipping soda with 3 ex-girlfriends of my dad as he strolled in.

He gets real fuzzy on it all then but he remembers only that they all sat together and that he left with my mother after he bought them drinks and they socialized as a group.

That was it, as from then on "Mike and Jayne" were together for the next 57 years. My dad tells me the story of how they met, but I am stuck on just how the hell he can walk into a place where there are 3 of his previous girlfriends sitting there with his future wife, and not *one* of them sticks a knife in his ribs! Today that would be very rare I bet.

He had to have been charming and decent for that one night to be pulled off like that. Having money at such a young age from the farm being left to him from my grandfather, he was a catch back then I suppose, but that one thing showed me that he was indeed a good man.

Now, I am not saying the man was some hot hunk, I am just sharing how he was able to have quite the bachelor life at an early age while he still kept his mannerism in check. It fascinates me that my life is the continuum of Mike's. I am part of his own wild ride through life. I realize that he was not even 28 years of age by the time he had 6 kids and a wife to support, all while he was working 2 jobs to do this and I am stunned.

When I take into account all of the things that befell him since he met my mother, it is like my life makes sense coming from this journey which my father has made. It makes me think of the life of the famous author Alexander Dumas who wrote the book titled "The Count Of Monte Christo".

Alexander's father was a Captain for Napoleon and he was famous as he once killed a group of men and captured a massive fortune from a garrison in France in a war.

The end of the story plays out badly then, with Alexander's father dying in prison when Napoleon refused to later ransom him from jail. How could Alexander possibly *not* have lived a mad life when your father has been involved in such a whirlwind of events before him?

Poppy did not take up arms and he never was involved in courts of intrigue with kings like Napoleon. Yet he none the less took on the same monumental effort that he was destined to handle in just being my father. He is a man who went through so much more than I could ever capture in interviews over the phone I am learning. This man has watched every person around him go through woes and suffering while his own never mattered that much to him. He just lived through it and moved on. That is why I dedicate my work to him.

I am sorry for every "waster" day that I shamed myself in front of you, while I also forgive you for every beer can that I saw you holding in your hand. I am so glad that we are friends now at the end, and I love what you did for my family sir. I don't care for you to tell me that you love me aloud sir. I know it is not your way. I kind of earned that as well I guess.

I am just glad that I can use your life and all the many things which I saw you endure as my guidance now in my own life. How you held on while our family was shredded by the US justice system I will never know the extent of it's toll on you. I can only say that as I watched from my cell on Death Row, you were the one true beacon of my will to hold on to my strength. You had the guts to tell me to my face that I had fucked up, and that it was down to *me* alone to get my sorry ass out of there. Well said Poppy, and thank you for being honest when I needed it.

Well, I got out of jail and I made amends with mom before she died. I keep my promise to her daily to be a decent and kind man as well. I honor your life too, as I use what you endured previously to see myself as being "harder than life, and yet still kinder than love". You taught me that much.

Mr. Michael Yarris I am also dedicating this book to you because lets face it, you are one tough little bastard. You did it all with little education behind you. I love your wit, I love your determination to live on, even when all has been taken from you. I love you for being the one to give me my life, and I really respect how you loved my mother. Good for you dad, you got that one gift which we all want in life with her love. That is what we hope that each of us can have in our days. I understand why you live on now too...you are keeping Jayne alive by living despite pain or sorrow. You sit in that empty house just up the street from where you first met her and you keep thinking about her so she that is not yet quite dead.

I guess my dad knows a deep truth about dying. It it very simply how we really die twice when someone loves us: We die once upon our actual passing, and then we finally die when the last person to love us has also passed away. My father simply refuses to let my mother die in this way.

Poppy, my life like yours, has been a terrible burden on me sir. I guess all that I can do now is try my best to stay here for as long as I possibly can, only so that I can keep yours and moms memories both alive. I will hang in there so you are in my heart just that much longer...It's hard on me each day and at times I have wanted to quit, but I know now that I have to go on regardless of how hard life gets. I want to try and leave a message of my life as you have, one that is worthy of the time that it takes to read it all in words. If I do this well then I keep our message to life we each shared alive for others to follow sir. That is why I hope that you can excuse my having used your phone calls to me for this one part of my book.

So for you Michael Yarris Sr., I submit this work in your honor and memory. May anyone reading this book truly appreciate what a good man you tried to be.

    Respectfully, Your son, Nick

CONTENTS

	Acknowledgments	i
1	LIFE BROUGHT ME TO *HER*	1
2	JUST WHEN YOU THINK IT IS ALL CLEAR SKIES ABOVE YOU...	Pg 16
3	YOU GET WHAT YOU PAY FOR	Pg 31
4	THIS IS WONDERFUL OR I HAVE GONE *TOO* FAR!	Pg #42
5	FREE WATER AT THE CANADIAN BORDER!	Pg #52
6	ROASTED, BAKED AND THEN BURNT	Pg #71
7	ALL THE ELEMENTS	Pg #88

ACKNOWLEDGMENTS

I wish to thank each person who has ever let me down or hurt me
in life.
If not for you acts, I would now be just like who you are.

Thank you and good night.

1: LIFE BROUGHT ME TO *HER*

Laura said to me; "Nick, *please* get us out of here"...

Laying on the bed next to her at that moment I shrank inside from hearing those mournful words. Neither of us could sleep and it was nearly morning light now. We were on the second floor of an apartment complex that was located in Ilchester, England. It was March of 2017 and I was alone with Laura, the woman who is central to everything in this story. It had been a bad day and I knew it was going to be another.

You see, our newborn daughter Jaymie Leigh had died only recently. She was only 5 months old, the sweet little thing. I was the one who had put her down for a nap in this very room which we lay in at that time. It was also me who had gone back upstairs to look in on the baby, only to be hit by the worst moment of my life 30 minutes later. I was crushed as I walked back down those same stairs with my heart broken to bits because the baby was dead in my arms. When that happened in January of 2017 our lives crumbled in just moments it seemed like. It was still so raw then that I knew that nothing was going to help right away. I just held Laura gently as I had been doing all night long. You don't blink that kind of shit away, it takes time.

Jaymie Leigh was the child who had brought Laura and I together the year before, in 2016. That child was our promise of all these wonderful dreams and plans which we then had together. Good, endless *good* is what we felt that she represented to us as a couple.

Right then Laura was dealing with a daily ritual of going into our bedroom which we had shared with Jaymie Leigh, and she would be hit by the instant loss of her child's death repeatedly. It was worse when she would feel the need to go and check on a baby who was no longer there, and then she would melt in sorrow before me. I saw all of it and I held my own mental meltdown in check out of respect for this lovely woman's loss.

The night before was also a bad one as it usually was for us because we had the nightly battle to find any sleep that we could. We worried over how we would make it now that everything seemed so bleak. It was all so much of a struggle then, as we tried in any way that we could to get money for food, heat, or electricity on top of grief pressurizing us.

You see, I had nothing left financially after my previous wife walked out on me in one of the lamest acts of someone being ungrateful to me ever could perform. She even stole the last of my money cheaply, so that when I met Laura I truly had nothing. In fact, it was Laura who took me into her home and cared for me is how it went...

I cannot work physically as I have too many crippling injuries to do manual labor. I do not have one certificate of educational achievement. I am unqualified, or too old for most work.

I am a writer and a public speaker by trade...which basically is a true statement. In reality though, I sell "my story", or stories about my life for a way to survive. I don't get caught up in seeing myself as any sort of celebrity because it connects me to the media by doing this. It's just what I have to do.

I know that I took my own personal history of pitfalls and hardship that were handed to me, and I turned it into my craft or professional way to earn a living.

What does any of that mean?

Some describe me as being "A former Death Row prisoner". There is a whole long story of how that came to be. None is really relevant here.

In short, I am one of only 120 men in the world that was shown to be proven innocent by DNA evidence from America's Death Row. What I did with this event of my once being on Death Row wrongfully, is to become a best selling author, and one of the best public speakers of my genre. It has taken me 13 years since my release in 2004 to have gotten to where I am now. Getting out of prison was the start of my life, and as such, it has been just as hard as when I was still inside of a prisoner's cell.

After being released from serving 23 years in solitary confinement, I had an enormous task just to handle the many hard facets of actual freedom. My limited ability to earn money has always been a struggle on top of whatever I face.

Still I managed use my education from my prison time cleverly I thought, when I wrote my first book for a major publisher. That first effort to tell my prison story was titled 'Seven Days To Live' which was published in 2007 by Harper-Collins UK. (Before things gets complicated, that book would be later sabotaged and canceled).

Following this I then became the star of my very own "docu-drama" film titled "Fear Of 13".

This film would take 8 years to be released, but it is now pretty much my calling card to the digital world. This film is very much central to this story of my meeting Laura as it is why I was with her back in the UK to begin with.

Basically just to explain things for all, in this film documentary which I mention, I told my own story in the best way that I could as an interview about prison life. Then someone else made it into a film/drama in a very non linear storytelling way. They tried to play on whatever twists that could be gotten from my tale and so it goes... Right as I begin this book it now airs on "Netflix" or elsewhere globally.

It was this very film being shown on BBC TV the year prior, in January of 2016, that had spun my whole world towards my being in that bed that I was sharing with Laura in another non linear way...

Back then I was carrying a whole lot of hurt in January of 2016. I was homeless and living in a crappy 1985 recreational vehicle that was mostly parked in Venice Beach California during those days. As I wrote previously, my ex-wife left me and I was actually left caring for two small dogs we owned together while feeling really low. I was feeling as badly kicked in the nuts as ever could be felt for a man, while I also felt like all my efforts to gain a stable life were over.

When my phone went crazy with thousands of messages from viewers of my film back in the UK who had just watched it on national TV, I was so not ready for any of it occurring. I was actually sick and weak from flu while I was bitterly not caring for myself, all due to a badly bruised ego of mine I confess.

I got hit with 20.000 messages alone the first day the film was shown. It was not just having my current wife then walk out on me for some guy whom she met in a bar that made me feel really gutted just then, it was even deeper than that.

My ex-wife of many years ago is a determined stalker in my life even today, and she was making my life prickly by obsessively trying to ruin any chances I had from my film being shown.

When my film "The Fear Of 13" was shown to over 600.000 viewers the first night that it aired alone, she went full mental outburst. Her new husband was even helping her this time no less, as they were trying hard to tarnish my image and bashing me on the social media site, "Facebook". That got under my skin and it was motivation enough to make me act. I set up plans to drop my dogs off and heal as best that I could. I was determined to get my ass to a hospital for treatment and find a way to then get on a plane back to the UK.

I saw from these many messages that I could once again try to hustle my story and change the conditions that I was in. Too many offers for stage plays, TV shows and such were pouring in for me to ignore that chance given to me. I knew that I had to get out of a dingy RV, wipe my sad tears and get my ass in gear.

A guy who saw the film that lives in Wales by the name of Nick Pritchard sent me money for an airline ticket without asking that a penny be paid back. Suddenly there I was in London, chasing offers for plays, or to meet a publishing house so that my first book "Seven Days To Live" could now re-published by Random House publishers under a new title. For weeks I drove all over the UK trying to secure anything I was offered.

That is when I met Laura. It was in a place called Basingstoke England, just shortly after I got there in 2016.

I could not have been more of a dickhead to her at first. Oh my shitty little ego was blazing angry still and I was not going to be one to have anything to do with another long term relationship and blah, blah, blah I told her!

Everyone is entitled to be a dick at some point and I picked that one day to be one is how I am saying it now.

Anyone who gets me knows that all along, I just want to be a dad and to be a family man. I just had to get over myself and let this woman have a chance. I met her at a time when I was really not sure of myself and it showed.

At the time that we each met one another, Laura and I both had gone through a whole lot of hardships in life separately. We were not love struck teens who were looking for passion with our heads in the clouds. We were each at that point where we just wanted someone of their word to show us the respect which we wanted. Her life just before me had left her with a situation that a lot of women have to face in life as a single mother. She had two children and was caring for them full time. The rest really does not matter as it is all just what she and I know now of one another.

That is how we got together and what I was doing there in the UK at the time we began. Now back to that morning in Ilchester, ...

Right at the very moment that I was holding Laura in that bed which we shared, all of the things leading up to my getting together with her there paled.

I just considered over and over how desperately this woman needed me to come up with a solution to our woes.

Looking at my options before me, I knew that I needed a whole lot of luck to bring about any sort of real changes. I knew only one thing was true if I failed to do this: If we stayed in that flat that we shared, we were doomed. If we broke down I was going to leave her as well as her two other surviving children scarred if I bailed out.

I knew that her daughter Zara Rose, who was barely 3 years old then, was just bonding well with me and really needed a daddy. So too was her oldest daughter named Bethany Grace, who at the age 6 was dealing with a whole lot more because the death of Jaymie Leigh was more real to her.

I was so in love with them each, that the thought of my not being able to help them in life just stole all of my joy away. It was a bleak time and yet we all tried so hard to live with things as they were forced upon us.

No one can possibly know what it is like to be me. And in those days in which this story begins, from the outside I was that oddity that everyone in this small community knew about but hardly knew me.

I was known as "The American man from death row", or "The American bloke who was in prison", or finally I was also said to be "The American man who was on Death Row who now has that woman with the dead baby"...

It got worse in time. Some of the gossip mongers went around the little school where Bethany and Zara attended class with whispers of what they thought had happened to Jaymie Leigh.

Surely the weirdo from death row must have done some evil shit to her they said behind my back!

The local school totally vouched for Laura and their testifying to the caring love which they saw Laura give to all of her children really helped. Still, the fact that parents were talking in front of children at home and also saying horrible shit behind our backs aloud made it hard to face down without responding. I made sure in every way possible to go to the school and not be affected by the "Chinese whisperers".

Each time I went to drop off or pick up Zara and Beth at school I went out of my way to be overly gregarious and charming with the mothers who were close to Laura, while I looked all others squarely in the face hoping they had something to say. Otherwise I kept well away from all the ones whom I knew were two faced behind my back.

Even the police had a low moment by playing with me mentally in a move that really hurt Laura and I both.

It happened shortly after Jaymie had died. It was when the police came to our flat early the following evening. It was arranged for them to visit us after they called us on Laura's phone and their telling Laura they had important "news" about her child's death.

As they drove over to our place to inform us we were on pins and needles hoping that we could find out what had caused a baby's death.

It was soon after they got there that we found out that this was all a ruse.

We saw this play out when the two police officers who showed up explained that they really were there to point out that I had posted a "Tweet" on the social media called "Twitter" about the preciousness of life.

The post in question was time- stamped in the internet log at being posted before the incident occurred that day. I had posted this Tweet in the midst of my grief, just hours after the death of this newborn. I posted this lovely message about a reminder of holding onto others in life is all. I used my American phone that was linked to California time to post this message. I knew the answer was simple enough.

I left my phone stay linked to California time while I was in England as I was constantly on the phone to my friend Anthony Samadani, one of then men who is producing my major motion picture deal that is based on my first book.

The play by the police was this...see if they could get me to confess to harming a baby by pointing out that my posted "tweet time" was off on the day she had died. They wanted to see if they could catch me off guard with this questioning when they said falsely that they were there with "news" for us.

They told us in defense of this ploy being used on us right then, by saying that some anonymous person had contacted them about my social media posting. It was they who had asked the police how I could post something online about my child dying a half hour before it even had happened.

My reaction to this attempt to dupe me into a moment as I said was not even to react. I knew instantly to not get angry or react in a hostile manner as I began to walk right into the kitchen and hand them my phone which was still set to California time.

Upon swiping the screen of the phone to open it in front of them right there, they saw that the phone was indeed set 8 hours earlier than the current UK time was right at that moment.

I even told them that I understood that they have to look at all things when investigating a child dying suddenly, and I tried to stay as calm as I could to show empathy for how their job must be hard on them to have to hurt me like that.

Meanwhile Laura went off on them and rightly so. She had no time for this bullshit at all. She and I were having a nap together on the sofa downstairs when our child died in the bed above us.

That is the sum total of events in one very bad half hour in our lives. It was like accusing her of being in on the killing of her own child. I tried to calm her down and I pleaded with her to ignore their actions. But Laura flew into them and told both a male and a female officer how shameful it was that they used a guise of helping us with news about the death of our baby, only to trick us like they had. Laura said this was so reckless and cruel by pointing out that we had company in our flat right then, and how thankful she was that it was her close friend Taz Middleton visiting. Thank God it actually was "Taz" who was visiting and not some person who could be a gossip with this incident later on. I just held my cool all the way through this shit because in their eyes, I was not going to be given the slightest break about what my mental state was by the police.

I tried really hard in the days after the police incident to not mess my own head up over how insultingly I was approached like I was by them. I hated that feeling that I was having inside.

They were making me feel like they thought that just because I was in prison among the "monsters" who hurt kids, that I had somehow lost my soul too. I am not broken and my experiences in life made me have a sweet nature and kind mannered, not mean.

To this day, I do not know how I did not get angry at the police when I could have exploded in reaction. I get all emotional at times and I am choked up now thinking of how I should have tried to defended myself. When they did this to me it hurt deeply.

I was made to feel ashamed for a blink of time and it bothered me deeply to think people wondered if I could do that to a baby. Eventually though, I only wind up feeling stupefied in the end.

I realize that no matter what I had said to them back then at that moment, it doesn't really matter now. You see, Jaymie Leigh is dead and that's not going to change. I loved her from birth to death and she never even heard my raised voice once. I live with knowing I was nothing but good to her.

Look, sad about it or not, I am just going to be stuck at times being viewed in my life as someone who just *has* to be twisted in my brain. I am faced with this situation of being thought of as a "nut", or that I must be broken mentally somehow for handling things in my life that very few can imagine themselves handling in theirs. When I am faced with this event happening to me, I do one really important thing to myself:

I tell myself that I am a better human being than they are being to me right then. I reinforce how I would be a lot more respectful to them than they are able to show me at all. I then tell myself that you can never defend yourself in life. You just are who you are, and time shows that to be so to others.

No, it was not the people talking about me at school or behind my back, nor was it the police wondering if I had done something to this child that was hard for me. It was a lot deeper than that. I had to stop feeling like if only I had somehow gotten things done faster, or if only we had gotten to go sooner.

It was my feelings of how I brought my own bad luck to this person's life and my terrible fortune in life was the reason all of it happened. Like I am one big curse upon whoever loves me.

All I am able to do is humbly try to make amends if this is so. I can only try to be the sweetest gift to Laura and hope it is enough. I have to only be somehow worth it all.

Now, I concede that I have seen a *lot* of hard things in my life both in and outside of a prison. I have a "list", or a small portfolio of images from these events in my life that I will never let go of in my memories.

We all have this thing in our heads like that. I get it that on a mental level quite complex, how some things are seemingly carved into our memory so deeply that we can never let them blur or fade with time. My own are widely ranging both good or bad. Things like:

I can remember Patterson Elementary school and the time I won an award for my school project at age 9.

I can tell you about the smell of an insane murderer I lived next to on Death Row who emitted ammonia odors from his flesh.

I can tell you what a soft pretzel coated with yellow mustard and lemon flavored Italian shaved ice tastes like on a summer day in Philadelphia.

Or, I can tell you exactly what it feels like to be stabbed with a blunt edged home-made knife in a prison cell, or to have a clean slash from a sharp bladed razor that went across my neck from an attack that left me with a rip in my skin long enough for 86 stitches.

I can even tell you about the time when my older brother Mikey and I as boys, had carved our name in the new cement of the sidewalk. It was right there on the corner or Elmwood avenue, just by our house back in the summer of 1974 that is still there to this day.

I can sadly tell you what it is like to carry a dead baby down the steps of your home to your woman who is screaming in terror as you approach her, as the child that you hold in your arms is dead.

I get so fucked up in my dreams over that one moment. I go through it over and over, and I cannot help it that no matter what I want, it never changes for me.. I look down at Jaymie and her eyes are not looking back. I have the same nightmare occur every other night when I get finally to fall asleep. Only thing is, I am not able to stop feeling this child's presence in my arms-even now. It happens over and over and if I just put my arms outwards with my eyes closed, there will be that weight of her in my grasp. If somehow it could be measured, I know it is real.

I go through this thing in my head over too, how I was the first person to hold this child when she was born into this world (as Laura was doped up in surgery from the birth and they handed her to me first). And, I was sadly also the last human to hold her in this life, as it was I who had been the one to lay her little body down for that fateful nap that she never awoke from...

I just wish that I had some way to deal with making sure this one bad image from my catalog of bad memories that comes into my head does not break me down. I am trying so hard now as I write this book, to put my own issues secondly, as the *real* test is the one that Laura is handling each day.

I won't let my own issues ever overshadow anyone else in life. I am not a drama queen like that in my make up as a man. But it fuckin' hurts to deal with so much that is attached to this child's death for me. Then there is the never ending battle of how do I figure a way out of it all for this new family of mine to survive.

I had to figure out a way to get Laura and I, as well as the two surviving children all out of Ilchester fast. I was racing against a very vicious time frame that was seemingly relentless in our punishment daily. I knew that if I did not get us away from there and try to change the daily ritual that was sucking us dry, we were indeed going to be broken by it. Even a street fighter like me knows that when the odds are hopelessly against you, you don't lie to yourself about it.

Of course what with this being my life, and also what with me having no money at all on top of being in debt to others financially when Jaymie died, I had only one hope of using my wits to overcome what seemed impossible.

I had to make up a good game and play it well or this was all doom and gloom. Life ain't nothing but games and I am one fine chess player. Something clever was called for and I was hoping once again that I was about to pull off another comeback. How was I going to make up a game?

Well, we all do this in a way actually. It's called trying to find a way to live. With no hope of physical jobs available to me, I was going to have to try this in another form.

It all began with a piece of paper and a single British Pound coin actually...

2. JUST WHEN YOU THINK IT IS ALL CLEAR SKIES ABOVE YOU…

What a shame that what was going so well for me and Laura had all turned so bleak once again for us both. The truth is that despite having no money, Laura and I were really happy in our first year together. Living in a small village that was located in the west country of England, things there are quiet and quaintly sweet most days. When I first got there I felt like I was so completely removed from Los Angeles life with its' bustle of 10 million people, that I could breathe again. It made me want to rest up and heal like I had been in a loud rock concert too long.

One of my best friends in life is a man named Marcin Puchowski. He is a man whom I used to work with while I was employed by a delivery company located back in Peterborough from 2010 to 2013. Whenever I come to England he makes sure I have whatever I need.

Marcin gave me a car to use while in the UK. He also got me a place to stay while there initially. It was just a cheap little Ford coupe and a real clanger that he gave me but it got me to where I needed to be. It was a rickety little thing that I had affectionately nicknamed "dizzy", because of the registration number of the car license plate beginning with DYZ. One of the doors was bent back, so the air rushed in and the car was loud as hell from the wind outside when you went over 50 mph. It was slow, ugly and free to drive.

Laura and I were happy just to have a car to go into town with or just to drive around together in the countryside while the kids were in school during the week.

We really spent so much time together that we were constantly growing closer. Those first few months were so sweet for us back then, as we really did all of the best things that you can do when you have very little: We cherished everything.

For us, we knew it was real simple to understand. We knew that when two people like us come together in life, ones who are so alike, it is explosive, challenging and rewarding all at once. She might only be a smidgen over Five feet tall, but my goodness she is a handful of womanhood. I met my match in so many cute intellectual and imaginative ways with Laura, that we are never bored. No matter where we are, it really is just us there and the world is this place around us that we step into and out of while swaying to our own rhythm.

I even began writing again at the encouraging of Laura. I wrote a stage play based on two men's story that was part of the film "Fear Of 13" that I titled "Big House Voices". I wrote a new book titled "Monsters and Madmen" about my prison ordeal that I had kept back from sharing previously. I have since followed that work with another book in 2017 which I wrote then that is titled "The Kindness Approach".

The point was that I was producing art again and it made me so happy that Laura was so good for me in this way. She devoured my work and pushed me to finish whatever I started. If I wrote a work in 3 days, she read it in 3 hours.

I finally had someone who was making me want to just be creative and show her how much I could make in art, if I just had someone like HER to believe in me.

With my writing under way again I began working with 3 men in the UK who are the owners of a company they formed which is called "3-Point Plan".

A writer, (Jivi Singh) A producer, (Nick Pitt) as well as a director, (David Evans) are the team I now work with. They affectionately call me the fourth part of 3-Point Plan because they have made me that much of a partner.

Back when my film documentary "Fear Of 13" was broadcast on national TV in the UK, it was these three men who decided that they were going to see to it that I got a good opportunity from my life story. I met with the three of them in London in February 2016 and I laid out for them the plan I wanted to make come about for me.

I pointed out the very obvious "big" thing which I needed to do with the help of these three men when I first met them. I told them that I needed to create a set of "life-rights" on paper that could be sold first . I knew enough to know how I needed to come up with a way to have someone *purchase* my rights a second time for more money once someone "owned" my rights.

In a complex way I had to have them further sell those rights as material for a TV show, major motion picture, and or radio shows to networks individually. That is the maximum way to get paid.

I explained that a savvy person could sign two separate deals in a very clever way if two different sets of producers were making separate projects.

I did not have the second set of producers then but I was planning on them coming to me soon enough. No way could I have known as I sat there telling these three men what I needed to concoct in order to get my life moving forward, that they were all for it. Not only would these men help me get my wishes granted, they also were bent on showing me how they were going to protect me in each deal.

I knew that unlike other production teams from my past, this was all a new set of people in their make up and character, so I had to trust them even if I did not already love them personally.

Thus, I sold my TV rights to these really nice men for one British Pound coin. I then sold them my motion picture film rights on top of that deal for another single British Penny.

All I had to do next was wait for a motion picture company to come along in the future and want these rights. I would have them in writing with a real production company to offer them exactly this later on. That way, there was a better value as someone else had already acquired my rights. I learned years ago that someone has to create a "piece of paper" that then becomes the actual bartering material for your rights to be sold.

Jivi, David, Nick and I then got underway with the making of the pre-production package of a TV show which will be titled "Dead Man Talking". I did so by working with the main writer Jivi Singh endlessly and then making up a video trailer and finally the first couple of episodes written.

As the main writer of this new material we were making, Jivi and I basically spent way too much time talking each day. It got deeper as I made voice recordings on top of these while out on walks near Laura's home.

I would walk along the most beautiful countryside while orally recalling the most brutal incidents from my prison life. In time though I had given this one writer more than I had even done before. He really is one of my most intimate male friends and I consider him my spiritual guide of sorts.

Random House publishers agreed to give me a modest advance for the right to re-publish my first book in 2016 from my efforts with them.

I then agreed to let them change the title of my first book to whatever they wanted it to be. Their bright idea was to change the name of "Seven Days To Live" to make it the same as the Netflix documentary title "Fear Of 13".

I saw what was at play in that my book was just going to be squashed into whatever they felt was their way of marketing it now based on my film's popularity. I just hoped it led to my getting booked as a speaker again in schools or businesses.

I managed to snag a few chances to speak briefly, thanks to friends who are educators, people like my close friend Emma Dobson. Work was scarce otherwise, and I had to play the waiting game in my life once more.

When the film "Fear Of 13" went out on the world wide broadcast network "Netflix" in June 2016, the very same endless amounts of offers came in on my social media. In fact that was how I met Anthony Samadani.

As a film and TV producer in the USA this man saw the value of what my story could be in a major motion film. He also saw how my story could be sold in many other aspects.

It was Anthony who had gotten me accept him representing my film rights in Hollywood. I put him in touch with Jivi and the boys back in the UK so both sides could sort out the cutting up of my rights. Within months I was having a real shot set up for my ability to land a major motion picture deal with Anthony.

I really pushed myself to keep writing then, I was also hoping to keep trying to promote the book being set for release, regardless of the new title it now was under. Otherwise I had to hod tight and to enjoy all that I had with having a home again and food to eat provided to me by Laura.

As all of this was going on all around us, Laura was happily pregnant with Jaymie Leigh. She was glowing and smiling and we were in love so much so that the baby was prospering inside of Laura.

We were preparing for this child to come into this world with all of these hopes and dreams that my film deal and or my TV, or even my writing would allow us all to move on...

On August 11th 2016 Jaymie Leigh Yarris was born in Yeovil hospital and it all seemed so wonderful. The love affair I had been in with Laura made her birthing such a breeze. In no time this child was full of joy and sweet days surrounded her in her mother's arms.

Even Laura's family was happy to see that I was so involved in caring for all of Laura's children, while I was clearly caring for Laura as well. That is all any family wants. They seemed even to accept my odd life. That's always hardest I guess.

To say that I have a strange life that I live as compared to most people is really putting things mildly. I have a pseudo-celebrity sort of life if you can call it this. With my book that is now titled "Fear Of 13, Countdown To Execution" being part of a major publisher's portfolio, I was asked to go on a press tour in the UK to promote it. Soon Millions of UK residents heard me on television or radio speaking about my life and my book. When I did press live on TV it was like some performance trip. All along I have no choice but to be utterly sincere with my reactions no matter how emotional it is to do so. It is so mentally hard on me to open up like this, yet I seem somehow to do extraordinary interviews.

My face was constantly on the pages of newspapers all over London and elsewhere in the UK in this way.

I ended up going to all sorts of places from this press junket in addition to just press opportunities. I even got invited to the Hay-on-Wye Book festival in Wales just before my book release date in October of 2016.

Laura and I had this magical trip through the mountains of Wales while Jaymie Leigh was with her sisters and grandmother named Jean Marie, having a good time back in Bristol England. I swear it was like some romantic film we were allowed to be part of as it all unfolded. We stayed in a Georgian manner converted to a hotel, we strolled along little shops and villages in the rain feeling like we were on some magical adventure all our own making.

After I did my book signing and I received what was said to be the only standing ovation ever, we walked around this festive little town where every other storefront was a book shop.

We held hands while we walked as I told Laura all of my dreams for what we could have together. We even carved our name on a bench by the river while we told ourselves how this was going to be forever.

When I got invited by the European Union in October of 2016 to come to Geneva as a special guest speaker before the United Nations, I knew that would be the transitioning point for both Laura and I.

The part of this story that is mine alone I guess, is how I had to watch another person begin to handle what being central in my life is to them. For whatever reason, it seems like Laura had to face going through having women in her life measure her in this weird way right before me. A few of the crass ones even had the nerve to ask Laura what was so different about her that she could be with me.

In their cruelty it is as if 'they' have to know what would make someone like me, (in their mind, for they have no clue what I am about anyway), would want to be with a single mother like Laura who has children. What is so unique about her that she can have this amazing life now being portrayed in the press, or how can she have a love affair that they can not imagine having happen for them? How can she handle such things as being on a world stage of events with me when she is just this average woman from Ilchester?

I knew that I had to show Laura how she was indeed able to be part and parcel to all of these things. I had to do so without any of those such thoughts or words from others bothering her.

I knew what to do in terms of showing her how to have an honest self evaluation of herself, and it was easily done with my own life as an example to her.

I mean come on folks, I am just a street kid from Philly who got this amazing chance to erase all that was broken inside merely with one trip to the death house. If I could transition my life from there and then, to being invited before world bodies and audiences of huge importance, then so too should she see her ability to be right there beside me.

That's what I did for Laura and I both. Thank God I had the sense and a special chance to bring her with me to places like Germany, Ireland, Wales and to Geneva Switzerland.

We were having this wonderful romantically international gala of a time due to this as we met all so many people together on our trips. Each time that someone had the delight of meeting Laura, they found her to be everything that I could hope for in a partner.

I knew surely that once I had stood before the UN while I made sure Laura was respected and accepted by so many in our dealings there, that would be the answer. I took her to Lake Geneva with a purpose and a plan for us that way.

With Jaymie Leigh comfortably in a baby carriage being pushed along by me in front of us as we strolled along this beautiful lakefront in Geneva, I knew what we had to do that night. I had to lay all of it out to Laura then and there so we could move further.

Laura and I love trees. We each have this appreciation for how old they can be. We like it how they have such a life span, that time can be measured by humans.

As we strolled along the side of Lake Geneva together we looked at the spotlighted tress before the buildings and were in awe of them. I told her how proud I was that she took this trip with me as it had allowed her to show to her own self that she belongs with me.

The women of the EU and the UN whom we met all felt such and instant like for Laura. We had a lovely time with the staff of this organizations who were cooing over Jaymie Leigh. I pointed out to Laura that she really needed to never let anyone make her feel like she could not change her life for one far flung from one she was expected to have.

It was in front of one particularly beautiful tree on lake Geneva that we stopped on our walk that night. Weeks of my thinking about it all had led me to that one point in time. I decided then and there we were going to set out things so that we were ready for that next part of our lives together. We needed no one else and Jaymie being beside us was enough for Laura and I to complete it.

There standing before me, bathed in the sparkling lights from thousands of lights around the lake, was my Laura. As I looked at her standing before me I kept thinking about how she had overcome so many factors in her life before my time in it. So many of her challenges were also my own experiences in life to deal with that I identified with who she wanted to be seen as.

Surely it was that one truism that those of us who share life experiences, make closer bonds. I also knew that within her was someone who was so ready to finally be free.

Free. Yeah, that is exactly what I saw that she finally is able to be with me

Free to be her open, honestly laughing and talking animatedly about life persona, all without a single fear of being shut down for it. I get to be with the version of her who is not being made to withhold her own personality because she is not allowed to be what she always felt she could be for a man.

Free to feel sexy and wanted and desired. Free to be listened to, because her words mattered. That kind of freedom.

I could not have been more proud of her right then and there. Just to know she was so grateful for our love affair made me want to set in stone how we would now go from here. She had trusted me to come into her life and to finally be someone she was not bullied by. I understood that feeling from all the foot prints that were left on my chest from those who had tread on me in passing. I am not like that. I want whomever I am lucky enough to be with to be completely familiar with me. I want my intimacy with Laura linked well beyond our amazingly endless physically entwined lovemaking.

I wanted her and I to be so in unison in every way that she never felt once that she had to shy away from expressing herself to me. I know that when I was lacking in my ability to feel such ways in my past how it kept me from really ever being committed to that person. I had a complete appreciation for my gift to her in making her a free person. With a sly smile I said "Let's get married right here tonight"!

I knew she was going to either laugh at the joke of it all- of how it does not work like that, or that she was going to wait to see where I was going with this...

So before she got into all that, I just grabbed her up in my arms and I told her how we should just do it all our selves ,right there, that evening. Let's just stand here before this amazing tree and we can tell each other whatever we want, I said. At the end of it we will swear before this tree how we won't end our bond that way.

Like two children who came up with a naughty game we wanted to play we decided that we were going to have "Our Moment". It was going to be that night. We wanted nothing to do with preachers, wedding cake or boozed up people dancing for us. Nothing to do with paperwork, bridesmaids, and a big deal honeymoon trip.

Instead what we would do is pour ourselves out to one another and state then what we wanted for our future. We tenderly sat with the baby beside us and we held one another. What she said was honest and truthfully apparent. Just wanted more of the good we had and for me to not be with others and just give her a chance to be enough of an adult to be happy.

I lit her heart up with all the many words which I spoke to Laura that night. I truly can emote the best honest feelings when my heart is opened up. I know it. I can just burst for how much tenderness it all is for me. I know I am rare for having the ability to do this despite a harder exterior shown to the world. I do. I have an amazing gift for my being one of the finest romantic examples of a man. I enjoy knowing my mother would be so proud of me for it as well.

As I stood there before her I was dressed in a beautiful suit. I was having this glow of the energy pumping through me from the crowd of people who had just heard me speak.

   Speaking before a huge crowd is energizing emotionally uplifting event. I was every bit in my element and feeling like all the world's kinetic force was pumping through my veins. There are just brief moments when I can feel like a beautiful man in life and that night was one of them.

   To her delight Laura saw me then as if was all things languid and sublime while she stood so feminine and alive before me. No matter what I write of it here, that incident was such a special moment we alone hold now in exclusive memory. It will always be one of those times that does not do it justice in mere words alone. One of those events that humans share in life, where just two sets of eyes created the only memories that could ever capture it properly.

   Standing there with her body pressed into mine as the chilly air was coming off of the lake made us shiver, We just did what we always do in such moments. We held each other and tried our best to hold onto every nuance of this really amazing experience which we had created. With nothing more than a smile and some really lovely words we got married that night.

   We always say that we want to go back to that tree by lake Geneva some day. We now say how we want to share some of Jaymie's ashes with that tree there as well. We want to show our own growth from that moment then, to when we next return. I do hope we get that chance somehow. I sure would like to have another chance to really appreciate who I was then at that moment to who I am now. I know Laura needs it.

Thank you Laura for that one amazing night. You never think that I am being corny or stupid for trying to finally be really romantically colorful.

I am not ashamed to let you hear me say the most delicate of words. You never make me feel like I have to work to share the ways in which I adore you. You really are all the best aspects of being a woman.

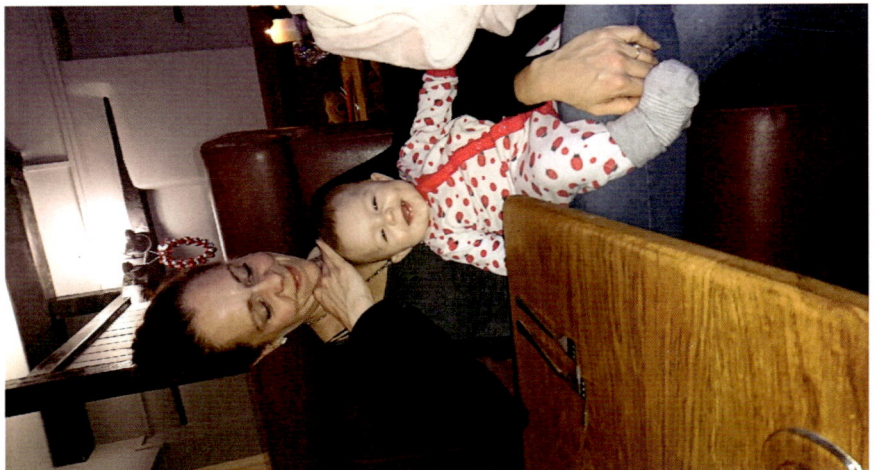

What can never be undone is what Jaymie leaves for us now...

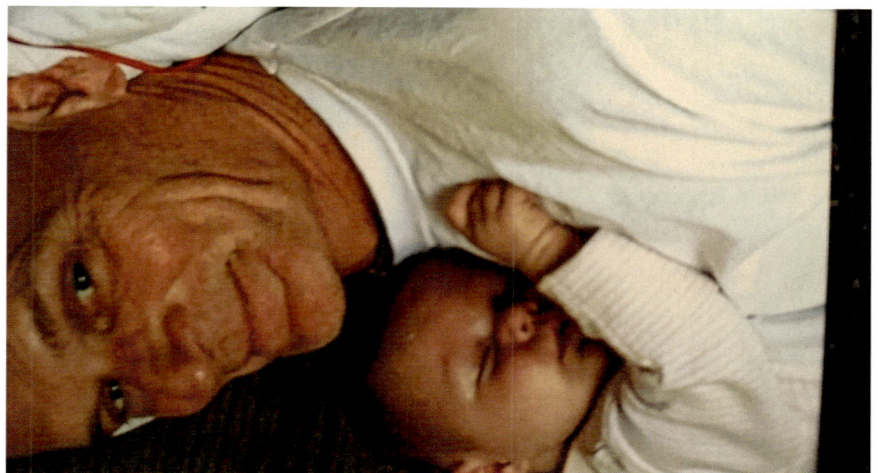

You will be always on my heart like the weight of love...

With an endlessness sky above we spoke such words of love...

3 YOU GET WHAT YOU PAY FOR

Christmas celebration 2016 was all about the children. Laura and I got them gifts with the little we had and we had a nice dinner at home and that was it really. We told ourselves that it really did not matter what we did in gifts because Jaymie was our gift and we had so many wonderful moments with her that year. We had so much hope and we were so sure that in the new year things would allow us to have better.

I hate this part. Gotta now go through the moments again of explaining how this all ended in January 2017 with the death of our child. I am hurrying through this just so we can segue into what we did from there is all. My happy times have all seemed so brief anyhow so please just let me get past this point and we can move forward. I am not trying to play on anyone's heartstrings.

All I know is that we got wiped out of whatever money we had to the point that we ended up borrowing money when burying Jaymie. Whatever wonderful times we had just got eviscerated and it was now down to me to save us somehow. Not understating and not making it worse than it was.

What I have always tried to do with grief is to make something good of whatever happened. I am not wired like others. I have this feeling inside that if I can do some act with my grief that I can than use what has befallen me to a better healing.

Just by happenstance a chance meeting back in December before Christmas became my chance. I met a man named Adrian Hindley in a town near our own through my friend Paul Beltrami, an artist who does sound recordings as a profession.

Paul told me that his friend Adrain needed help or he was not going to live much longer. Paul asked me to go meet this man and see if I could get him to stop drinking himself into oblivion.

So just before Christmas I go meet this man in his flat. He was disheveled and booze was reeking from his skin. He gyrated and swayed unsteadily on his feet as he did his poetry, all while hovering just inches from my face as I sat in a chair.

Now, unlike most people, I am used to dealing with humans who are unstable or worse. I never flinched as I let Adrian do all of his artful antics. I was fascinated to say the least. What a complex brain and yet so sad that he was swimming in booze.

I am not a miracle worker and I have no magical way to get others to change. In reality we all change in life for our own personal reasons. I just am glad that I can reach some people is what I like to say.

Adrian suffers from Bi-Polar-ism and this affliction is so hard on people once they drink booze. I saw that in a sober state he could possibly handle things. I let Adrian show me these games of his which he invented over many years. I saw that this was the one true love of his life. That was how I could reach him.

I spent a couple of hours telling this man how beautiful he could be if he quit poisoning himself. I explained my own troubles mentally and how I had found a way to not be consumed by things. I did not know if I would reach him back then in December, I just told him that if he was serious and he quit drinking that I would get my friends in the USA to help him.

I am proud to say that even at the time of writing this work in late 2017 Adrian has not had a single drink.

My pal Adrian even has a can of beer saved unopened in the fridge at his flat to remind him of how he was giving up everything for that one item.

I brokered a deal with Adrian about how I would use my skills as a writer to write out all of the games rules for his games and I would write the company bibliography as well. I would further use all of my connections in Los Angeles and the UK for him to get things going if he purchased a plane ticket for me to go to Los Angeles so I could get my deal done.

I knew that Adrian and I could use my taking him to Los Angeles for several good reasons in March of 2017. I could go try to get my film deal done as I said while I also set up a possible one-man stage play there, all thanks to Anthony Samadani getting me in touch with people who could do this. Adrian could meet my friend Adam Callinan in Manhattan beach California while we were there first. Adam is a man who owns a company which he created himself. I told Adrian how Adam had started his company Bottlekeeper.com from scratch, and how he could possibly help Adrian get his games company started the same way. I then said we would drive to San Francisco to meet a publisher I knew named Don Dehart who could print Adrian's' card game.

Going to Los Angeles and leaving Laura was brutally hard on me mentally, but I had to go see if I could meet one of the richest women from South America for film financing at the time, or if the actor Chris Pine would take the project on. Honestly I would go do whatever else Anthony could get set up for me because I had no choice simply put.

I even managed to get contacted by a comedy troupe that is called "The Upright Citizens Brigade" who wanted me to come and do an improv performance with them in L.A.. I thought this would be perfect for me to sell copies of my books that I still have in California. I could invite the stage people and show them my abilities as well.

Now I thought I had crazy things occur in my life overly so, but Adrian is not far behind me folks. Being afflicted with Bi-Polar-ism is not an easy thing to deal with when you are fully manic. My friend "Ade", as everyone calls him can recall stories of his times doing things like dressing in foil and a hat with big glasses and riding a sparkly bike around town as he wanted to hire himself as a moving advertisement. The image of this pasty skinned, tall lanky Englishman on a bike out of his mind and trying to sell his idea to passers by by shouting "ADZ" as if they would understand from this the company name he thought would be perfect. Don't ask what the ADZ means, I tried to scrub that part out of my brain long ago.

The best Adrian stories are innocuous and sweet. He is such a cad in how he acts or reacts to things. Things like how he awoke one day found that someone had taken a stencil of a rat and painted the image on the side of Ade's block of flats. In a hurry to go to work he really had little time to investigate. As he is pulling out of his parking spot in front of his place he sees a boy looking up at the graffiti on Ade's house and Ade asked him if he is a "Hood Rat". The boy looks confused and Adrian points to the Rat on the wall and asks the boy if he is a "Hood Rat" too.

The traffic light changes and Ade drives off feeling smug for chiding the boy.

But of course this is Adrian's life, so Ade makes the wrong turn on a one-way system of streets and ends up right back in front of his own building. He pulls up with the same boy still there from just 2 minutes earlier when the first encounter happened. What does Adrian do? He asks the boy *again* if if he liked that rat image on the wall and if the boy was a "Hood Rat"...

Of course the boy is truly stunned by all this and Adrian drives off and goes to work thinking he at least got some boy to feel his sarcasm if nothing else. And then of course this is where the story turns to the police and a boy reporting to them that some man tried to accost him on the way to school and then said Adrian tried to recruit him into a gang called..."The Hood Rats"!

They way he tells it is much better, but man I can tell you this just keeps happening to this man and it sure continued on our trip to Los Angeles.

With only a few months of sobriety Adrian was like Bambi on ice during those days for how erratic he was. His eyes were still messed up and his mental faculties were still shaky too, so it was quite the adventure taking him and this huge plastic carrier bag around that was full of his games.

He is so sweet of nature in his personal make up that Ade is like that sweet elderly uncle who you want to help fix his shirt and he misses things slightly and you let it slide. I love the guy and I am so proud of his willingness to trust in me, when some told him that I would just use him.

In truth Adrian asked me to have a part of his company and I told him that I would not do that. I said I would do everything I would for him, just loan me a bit of money and get me to Los Angeles so Laura and I could get out of Ilchester.

This trip to Los Angeles was as I had expected it would be. It was a mash between my crazy luck and Adrian's. I drove 1.500 miles while Adrian told me every conceivable tale of his life as many times that his sober mind could do it. We laughed, I cried at times and so did Adrian as we drove along listening to Michael Franti songs on the car radio system.

   The whole trip felt like to me that no one was getting it and that they were not hearing how important my helping Adrian was for me. I really think at times people can pat me on the head and not even try to pay heed to what I have offered up to them. It happens and I just let it be a lesson to my ego to not respond...after all it just my feelings about it and I could be wrong. Just saying that my efforts to get Adrian hooked up felt like it was not what I had hoped for.

   Of course no one else knew what it was that I needed personally to come out of all that this trip was for me. I even tried to use the performance with the comedy troupe to explain it as part of my speaking that night. Hell, I was hoping to just make sense of it by explaining things aloud.

   My way to deal with grief was to get Adrian to come back to life. I needed to get him to be able to have his dreams and make sure I took nothing from it all so that it was altruistic and fair. =

   I know this sounds odd but I needed to show myself what a really good man I am so that what happened with Jaymie is not something I deserve. Don't worry this shit won't make a bit of sense to you my reader. Just, to me this was my way to show that I am not some broken being in need of suffering. I also wanted to see if I could really use my skills to get another man to heal.

I figured that if nothing else came of it and I still got nothing for my efforts of going to Los Angeles, so be it. At least Ade now had his chance and I did it all with panache and grace.

I told the audience the night of the Improv performance basically everything that I just wrote before I then let the troupe members interview me so we could use what had happened to me and Ade on the trip as material for their skit performance.

What a riotous fun time it all turned into then. What an absolute joy for me to whip off my suit and join in having fun making light of both Adrian and myself.

I was so caught up in things that I forgot to mention the books which I had brought that evening and that they were for sale there. Everyone had to leave for the next performance to begin and I ended up doing this whole thing without any money to come home with from sales.

I even passed on the request by Anthony of my staying an extra day to meet this billionaire woman who might invest in my film and I went back to the UK feeling like I just blew it.

Even my arranged dinner with the actor Chris Pine went over like a lead balloon as I totally lost sight of being "source material". I made my dear friend Alejandro Monteverde look badly for how I spoke way too much about the film aspects and the scope of it. One of those times where I should have just kept my mouth shut and I really felt badly as I got back to Ilchester with nothing but bad news of my trip.

I paid physically from a wild jaunt around a familiar track back in Los Angeles and I got back feeling exactly as my mother said I would during times like this.

My mother always said: "Nicky, you *get* what you pay for" so many times as a boy. Now, that could be applied to not spending cheaply for the reward of cheap items that you bring home or such things like that. In this instance however, it really did mean that what I was doing to try and trade my grief for a good thing to grow out of it all was a price that only I could know the value of.

What really matters more than my disappointment or satisfaction with that effort then, is that I proved again to myself that I am indeed a good man in every way possible.

I needed that boost to my confidence after the police made me feel low over Jaymie's death, or all the others who were hurting me in life. I needed to go and do another truly wonderful thing for another man so that I could reinforce all that best reasons why I love the man whom I am in life. No one needs to get this about me but each time someone or some thing hurts me like I was hurting right then, I do something few can. I make another person's lief less harsh for how it makes me heal and feel good about myself.

The thing is, I am going to get what I pay for in loving acts and not deeds of destruction and as such it hurts a whole lot less when you can at least lay your head down at night for knowing you did your best to be kind and caring.

Still wasn't paying for food and such but I was so hoping that I did my karma good by just being good.

Toss it all away...

Gather all of your courage and go...

The moments a journey begins as another one ends...

4 THIS IS WONDERFUL OF I HAVE GONE TOO FAR!

Finally by May of 2017 my boy Anthony Samadani got us some big time investors to put the development money together for my major motion picture deal to be done. Wow. This guy pulled this off while keeping Alejandro Monteverde as director. That kind of thing hardly ever happens in Hollywood. I still get to keep my deal with the producers in the UK which allows me to keep hopeful of working that way.

With the advance paid to me I could immediately pay back the people whom we owed money to. That was the very first thing I wanted to do. Then Laura and I started taking Bethany and Zara on short trips in the UK while we got our visa's and plane tickets to the USA done. It all started to flow fast and suddenly we got a skip and had it delivered outside of the flat we lived in. We then tossed all the things away that were clutter and confining. There was this real energy of change coming and the clouds of sorrow were being blown clear. I could see that all of this change was exactly what Laura needed. It was all real and our praying and waiting were not some lame torment to us.

It all played out so that in the last days before our trip to Los Angeles, we were sewing up all the loose ends of things and it was all lined up for us to be going. I was so excited for this chance and the children had their heads in the clouds with thoughts of Disneyland and America.

On the night before we flew out Marcin and Joanna Puchowski drove down to Heathrow airport with their son Oliver with them. They spent some time with Laura and I having a nice meal together with our two girls. Marcin came to pick up the car he had given to me and it was like we were all part of this trip that way.

I love it that my brother Marcin came down and made sure that I was okay mentally. He saw what the past two years had done to me. Marcin especially had been so heartbroken over Jaymie Leigh passing away. I knew all along he was doing this thing to make sure I was going to be alright.

Marcin is one of those men who is such a fine example of what a good man should be. He always tells me "It is a pleasure to help you". That kind of friendship is rare in life and I am so grateful for what he means to me. I try very hard to treat people the way Marcin treats me. I knew as I was saying goodbye to him that if not for what he had done for Laura and I we would have not been standing there. Marcin will have a very big head from my compliments and he will also torment me endlessly for writing this squishy stuff about how I feel about my friend. Oh well, Marcin is always the second funniest man in the room whenever he and I get together, so there's always that.

As the Virgin Atlantic flight took off on an 11 hour flight from London to Los Angeles I knew that was only a fraction of the journey ahead for us. We were going to go on a mad jaunt soon as we landed to go get a car, hotel and things like that. In Los Angeles, traffic makes those simple things a major nightmare. If I was smart I would get whatever rest that I could on that plane trip because I knew we had a whole other day ahead when we landed in early afternoon on California time.

I really did not get any sound sleep on that flight. What I did though was do some really serious consideration to all the things that were to about to become a reality. I had to really be sharp to pull off a trip with two young girls, a new partner and then to have Adrian come back and join us all on this trip. I set that last bit up so that I could once again help Ade after the first one really did not produce much results. This was a beast of a trip ahead and I knew I had to have some serious amounts of luck to make it.

I called my friend Noah Vanbenschoten before I left England and I told him that we may need him to come to LAX airport and help us get a rental car. I had no USA bank account and renting a car demanded that I have a current bank card. Noah and I have been friends since 2014 when our mutual friend Jason Daily introduced us. Like Marcin, this man is truly good to me and has been one of the really good things that came from my time in L.A.

Once Noah helped us get a rental car we went to Anaheim and checked into a hotel near Disneyland. I think that was the last quiet day since then folks. Within days we were running around Los Angeles like someone shot us out of a cannon. I made the mistake of dropping Laura off in a supersized Walmart store with the girls that was as big as the village they grew up in. By the time I had parked our car Laura was standing there just inside the entrance with the girls having lost their minds and ran off. She had no idea what I had just shoved her into. I was told then by my very annoyed woman that I was never to leave her alone in a place that big with these two tornadoes of "want this!" and "need this!" that the girls were being then.

Even now I have to laugh at how overwhelming it is for people like Laura to come here and see how spoiled America is for variety. She could not stop her eyes from being drawn to how there is like 900 choices of just peanut butter alone. The children went wild with how many toys and sweets American kids have access to. It was like watching time travelers come to the new future and they just cannot believe what they were seeing.

I know that it seems like it was all so frivolous at the start, but there was a real plan I had being set out. I had been working for months to set up a way to accomplish the erasing of so much sorrow for our surroundings being tied to a child's death. I further had what I thought was unique way to also tie in a business opportunity with it all.

Anthony got a company in Canada to agree to hire me to go and do a week long speaking for their 2.500 employees in August of 2017.

This business was located on the Western portion of Canada, which meant that I could drive there from Los Angeles and bring books to sell all week. The company agreed to pay my expense for driving there and I thought this was going to be a wonderful chance to have our holiday trip tied to my going there for making money as well.

All I had to do was find a vehicle big enough for us all to take enough books to Canada so that we made good money out of driving 2.000 miles each way. It had to be something we could also sleep in, as the money we had to risk to get to Canada had to cover our hotel costs for 3 months. So, I found a 1997 34 foot long Fleetwood motor-home that was located near Anaheim. Guy said "everything works" and we trusted him.

This is where I messed up. I was trying too hard too fast to get things going. We only had a week in Anaheim and I wanted to get the right RV and in my haste I got the wrong one. The guy selling it to me only bought it and then re-sold it and he did nothing to see if it had any issues. He put me and my family in a 34 foot Motor-home that had a tire with a 6 inch manufacture defect hidden on one of them. It was hidden between two of the rear tires and I missed it.

Two days after we got the RV we were in a park at 9:pm at night in Irvine California picking up "Terrry" and "Gina", our new puppies that were poodle mix breed. This is how America is. You want things, there on the internet everything is ready to be picked up or dropped off 24 hours a day. We found the dogs online and went and bought them that night. It was so cool to have the girls freaking out over their new puppies and watching fireworks nightly in Anaheim. Suddenly all of the woes and torment and daily rituals of sorrow filled events were being wiped away by new events. I had the girls and Laura right where I wanted them to be, far away from hurt and changing daily.

I knew instinctively that the new puppies would replace the maternal giving Laura was cheated out of. I knew that her nurturing would kick in and these two tiny baby pups would get all of this love from her.

I also knew further how Beth and Zara each needed to give these tiny dogs love as a healthy distraction for them. This was so important despite the headache of having young dogs to be trained or cared for on the road. I knew that it was worth it if the psychological replacement took hold as I had hoped. I used my education to lay out an entire journey that was also laced with things that were designed to get this family back onto a path of being happy again. I did not care how many hours or miles it would take to get it done. I did not care what others thought of our trip either, as it was really only about us.

There was one more really important thing we had to next do that would become one biggest healing features for Laura personally. This is ultra personal for her so I am going to be very delicate here. I am not trying to make myself look genius or any of this by explaining that I had a plan I was pulling off for these three people in my life. Still, I am very clever for all of what I put into my education.

I especially understand humans from on a unique level from my life's experiences and my own education. I knew enough to know that Laura had to overcome a personal issue that a lot of women deal with. One of the things I wanted to do was see to it that Laura had surgery to fix her breasts.

No man can understand truly how much of a woman's femininity is connected to her breasts. With a woman recently losing a child this is especially damaging to her psyche. I knew that if I could get Laura to a clinic in Los Angeles that she would see from a professionals standpoint how this could be such a benefit to her.

I adore this woman so much that I wanted to help her do this without it being damaging or negative. I let her have complete say in whether or not to go through with it, but I damn sure got her to that clinic so she could decide fairly without me being in the room.

My goals were that I wanted to give Beth and Zara this wonderful set of memories to completely erase burying their sister in January.

I wanted to give Laura the chance to let go of all that tormented her and to finally feel empowered enough to be whomever she felt like being. I then needed to find some way to get all of this pain and anger and hurt out of my system. I knew I was taking on a beast of a challenge physically and that was okay with me. As long as Laura and I were not sitting in the flat back in Ilchester I would take my chances that what I set out to do would indeed be our healing. When you have been pushed by life to change all that you held onto you take with you all of your memories or scars. What you turn them into then defines who you were before it happened to you is what I believe.

No matter what else is ever said about me in life I can always point to how I have always done for others more than I did for myself. Not sure that means a whole lot in measuring what a man I am, I just know it feels better to be this type of person.

Throughout this entire trip it has never been about me and it never will be. My healing is what comes from my being so consistent and loving in another person's life. My healing comes from knowing that what I gave has never been second to what I got.

As I lay in a hotel one night in Anaheim thinking about how I had pulled off so much already in just days of being there I looked around me right then.

The two girls were playing happily with the pups on the floor of the hotel room, Laura was sitting on the bed smiling at it all and the TV was showing some stupid advertisement about penis pills to help guys my age get an erection. I turned to her and I said with the biggest smile on my face: "This is wonderful of I have gone too far"! We laughed at the self aimed sarcasm I had just used and yet what a mouthful I had just uttered folks. What a huge one it was indeed.

I told Laura that while we waited for Ade to fly over 10 days after we had landed we should take the RV up the California coast and "boondock" with the RV along the sea.

I can at least say that this was so fun at first...

As we set out up the coast in the RV with the dogs, girls and a new cactus propped up in front of Laura who was sitting in the front seat, we had our best days of the trip.

I gotta say I am pretty good at going from a tiny little car in the UK to driving a 34 foot long beast of a vehicle. The girls thought it was amazing that they had beds and a big TV to watch while we drove. The huge front window was a constant changing vista for Laura's eyes. I had the best feeling in the entire trip while we floated for a week around the coast discovering places to park or foods to eat. The girls were in a dream for all the places they went. I watched a whole week pass and not once did Laura have a melt down over grief stricken moments. She would be quiet and sad at times and I would try to shake her free from it's grip with jokes or kissing and holding her. I told her that she should brace for when she had a whole day pass without a tear, or then a week. I told her not to beat herself senselessly over being happy and how that is all ego. Still she struggles and still the scars need time. I thought about once Adrian was there I could have some additional help and this would be nice.

The sweetest part of that week for me was one night Laura and I were outside of the RV as the children were asleep inside of it. We were parked along a quiet road and we sat on the curb having cold glasses of iced tea. We sat looking at the stars over Ventura California and the heat was not yet ruining being out at night.

I looked at this woman as she sat with the sea behind her that was shimmering and crashing in on the shoreline. He skin was so shiny and smooth and her hair was still short and neat. If I had met her right then, I would want to know her ,is what I felt inside. I like that feeling even now. There is just something about this one woman.

I wished that so much had gone differently for us on this journey.

If all I had was that one night when we sat there with our hands connected, that one moment of pure love that I don't want to let go of. As I sat there I blotted out every bad thing before then and I was able to just be grateful that I was the "good" that had come for another person to be lifted by. I was *her* answered prayers and dreams of longing or wants to be met finally. As I considered all the many things which this woman had taken from her or done wrongly to her in life before my meeting her, I felt the utmost tenderness towards her. I was determined that no matter what it was that I faced from there on, that this one feeling which I had was going to be my motivation all along.

And with just like that, as with all of the other times that energy has propelled changes our way, so too were we soon swept up on a mad dash back through California and further, this time with Adrian.

With all that innocence allows they dream of "sweetie land"...

What we wanted and what we needed were such simple things.

My view has always been while having your back babe...

5. ROASTED, BAKED AND THEN BURNT

I should have known that even picking Adrian up at LAX was going to be the start of things. While back in Anaheim with Laura and the girls I rented a car from TORO, which is a new phone app system for personal car rentals. I used the car to take the girls to a water park while I had the RV in for servicing. I was really angry that the entire brake system had to be fixed and replaced and I had to order two tires.

Of course I could not find Adrian and it went on and on until finally I found him in the international arrivals area. He was so out of it from the plane ride and I was just as tired from a long day in the incredible heat of Los Angeles. By the time we got back to Anaheim it was 10;pm at night and Adrian was on English time and wide awake.

The girls know Adrian from our trips down to his home over the past year with Laura. He gets on well with them but Ade is not a babysitter or minder of any kind. Our girls are too high energy for him and he soon was going bonkers with them driving him crazy.

I saw right off that Laura and I could not ask Adrian to have the girls for a bit while we got a break and he was all about getting things done with his business. Laura was not happy to have a third wheel making our world a bit cramped. We all settled in for that awkward at times but overall sweet journey that went like 5 days too long ,is what I think.

Still this all put a monkey wrench in my plans that while Ade was here with us I could facilitate Laura having surgery and this really complicated things with everything. Meanwhile the heat in LA was relentless and climbing. My efforts to set up meetings for Adrian went south because everyone's schedule was off. I was grounded with the RV and we were thinking we had to get out of there.

Our first trip with Adrian we went to Soledad canyon and camped in the high desert mountains with the Coyotes howling for us at night. We celebrated Laura's birthday and really tried to enjoy things.

But the RV was acting up so back to Anaheim to see what was wrong. Once again the emergency brake system was not working and our stabilizers were not functioning. If we stopped any place we had to put wood or bricks behind the wheels so the RV did not roll away.

Back in Anaheim we still did not get a chance to go into Disneyland. We saw endless fireworks but not one day spent inside the park. This was wearing on Laura and she was really not happy that the girls were so close but having no chance at having such an experience as Mickey and Minnie.

I had to go to Claremont CA. to go get my divorce papers which had finally come through and so we all went back to where I used to live. I had to go back to go to get my new bank card and DMV mail at my friend Jason Daily's house.

The RV center let us drive out of there with a tire that had the big gash on it and sure enough that tire blew out at 60 mph on the 605 freeway in 100 degree heat.

My ears were ringing as I got the RV to a stop. My head was pounding, as my was heart beating like crazy. Laura was livid that this guy Frank whom we bought the RV off of had ripped us off. She was so angry that the RV place had let us drive out on a tire they knew was damaged, and she was rightly upset that this happened with her girls now having to hide under a tree by the roadside baking in the heat.

Another huge expense and another bad moment with more crazy times I did not need. I was fit to be tied as I got the RV towed back to Anaheim with the girls hidden in a big tow truck and me riding in the RV that was on a tow hook. I honestly felt like this was surely the end of this and we could just get on with things as we got our bad luck over with.

So, RV gets fixed a second time, with new tires replacing the old ones. Off we go to try and fix Adrian's meetings and get his game designs printed on boxes. I agreed to do his video for funding on the social media as we could not get anyone else to do it. With the heat being so bad then, we decided to go to Big Bear lake in California 7.000 feet above the Los Angeles basin and well above the 105 degree heat.

With time running out on his time there I did as best that I could to help Adrian before he left. The trip to Big Bear was actually really good and we had so much fun the last days there. I extended our time in a lakeside RV resort so we could just drive down the day Adrian had to leave from Lax that evening. Laura and I could then go and meet a very well respect surgeon we wanted her to see about corrective surgery after Adrian went.

As we drove down and down the smell of burning brakes filled the RV with black smoke. We tried several times to sop and wait for the brakes to cool. It was harrowing the further I went. My emergency braking went first. I had a series of twisting turns exposing us and as I pushed the brake pedal it went right to the floor.

I shoved the gear lever into low and held on as I watched the speed increase no matter how hard I pressed the brake pedal. By 5.000 feet we were done as I saw a big slip of road on my right. I jammed the emergency brake, hit the back up power brakes, and stomped the brake pedal as the RV slowly and finally stopped against the side of the mountain. I screamed at Ade to get out with rocks and get them under the tires as all of the brake fluid then boiled out of the master cylinder.

We had no reception on our phone so a passer by stopped and rang the police for us. Cars slowly pulled past us and we all stood outside the RV in the heat shaking and scared for how this thing just would not stop.

As a large stream of brake fluid ran out from under the bottom of the RV Adrian actually asked me if I was going to keep going so that he could get to LAX.

I looked at him like he was still wearing his frigging ADZ suit while crazily pedaling a bike around. I told him to take his crazy English ass over to the guy who had just stopped to call the police for us and get him to take him down the mountain. Suddenly my man who had to have me basically do any and everything for him snapped to it like he was James frigging Bond and he suddenly got a ride, then got another ride, and was at LAX with a cold drink while waiting on his flight.

Meanwhile my sorry ass was still bouncing around with Laura and the kids inside of the RV as a Mexican gentleman - tow truck operator, wondered just how far he could get the RV to tip before flipping over at like 60 mph. It was nearly as bad as the blown out tire as he was taking us down a very big mountain. Our screaming faces in his rear-view mirror only seemed to spur him on...Come on man, we were surely *not* yelling "GO"!,

I promise you sir, it was "NO"! No, please God, no, no more like that for me ever again.

Another tow and more money lost. I lost all confidence with the RV and seriously as much as I joke this event shook me. I had night sweats and bad dreams following this one. How the hell could I take this RV up to Canada when I cannot even trust it to not kill us all each time out?

Back to Anaheim, drop the RV let my friend Lionel help us. We said that was it. Let's take the girls to Disneyland and pretend we just did not just waste three weeks of our lives and all this money for something that we had no business doing. I was so willing to try and anything to shake this bad luck off that I agreed that we should indeed just have a blast.

The moment the girls got to see that Disneyland was not some empty promise they melted. Even Laura let go of the things she was really hurt over. I told Anthony that I had a moment on the mountain where I thought this was it. I am really confident and I really handle a lot. But that moment when I had nothing to stop a huge RV was daunting. The little girls screaming still haunts me and I cannot stop that queasy feeling at times when I think of how close we were.

My going into Disney then was exactly what I needed as I was taking too many shots emotionally. I was putting in 20 hour days and driving way too many hours. The heat is of California never ending and it's relentless pounding is horrible. I am telling you that this global warming is so real because in just 2 years I could tell the difference in temps in LA. Soon it will be uninhabitable for humans during the daylight. I honestly believe that.

As our time in Disney played out all of my plans for how the trip was to go got blown up with the RV. I had to change Vehicles and do this another way. Meanwhile I had to use the time we had left to get the surgery done so Laura had some time to heal before I drove her and the girls 2.000 miles to Canada.

I still thought this was all still doable and that as long as I made it to Canada that I could recoup the losses what with the company who hired me paying my expenses and my book sales.

Everyone told me Canada was going to screw me over and that they were cruel and mean at the border. I explained how I had been there twice previously and I had a history of honoring my entry with them. I managed to get security clearance from both the UN and EU and surely this aided with my efforts. Still though, I wrote to the Prime Minister of Canada and began an exchange of letters with his office as well.

Taking the RV to Woodland Hills we parked it at my friend Noah's house while Noah took me to a dealership so I could purchase a second hand car. I eventually got a 2006 SUV and got a tow bar made for it so I could tow a box of books behind us. I then took Laura to San Juan Capistrano California so she could meet her surgeon.

Adam Callinan and his wife Katie took the girls for a night and Laura and I set a date for her to have the procedure done. It was so fascinating to Laura that like buying puppies in a park at 9;pm at night from the back of a car, you could just go pay for a medical procedure and there is no waiting. I told her that in America you can have anything as long as you can pay for it, and it makes sense. Just like that she was recovering and I was driving beside her waking her up every fifteen seconds because she was in a nod from the anesthesia. I hated seeing her so deep under and I blasted music and kept yelling for her to keep awake.

A few days after the surgery Laura and I took the girls up to San Francisco so that she and I could get married. We let Laura heal while we stayed at the Pentaluna KOA campgrounds. I watched this woman go through all of this post surgery with no pain pills blurring her. Wow. I mean it when I say that she virtually took nothing the whole way and I am so proud of her. She would look on-line and read posts of other women saying they had managed their first meal a week after their surgery and laughed at how she had Chinese a few hours after hers and she has not stopped since.

I was just glad that there was no infections of other side effect that ruined the trip. We were all set to go find a preacher or person the state recognized as legally allowed to marry us in a place called "Nick's Cove" California. The girls would be there and we just wanted to have the legal process simple as we had already had our moment back in Geneva.

Then Anthony rings me and I tell him our plans and he tells me to get my ass back to LA as he can marry us. I was laughing that we had done the crazy drive from Los Angeles thinking we would do this all in San Francisco and suddenly we are packing the car and driving 500 miles back south.

But for me, and I mean this sincerely, I really wanted Anthony to marry us. It is not because he got us the film deal and everything like that. I just wanted him to share our most precious moment because of who he is.

The reason I love this man is because like me, he has shaped his whole life just as I have, around a promise that he made to his mother. Her passing only solidified within him his vow to be a man who uses his life for others to have better. I know he was heavily influenced by his mentor Muhammad Ali whom he befriended years ago. That piousness and that drive Ali had in life really impacted my friend Anthony Samadani and he tries really sincerely to honor Ali in this way.

Worn thin from weeks of driving and camping I started driving back to Los Angeles for a legal marriage ceremony to be done by Anthony with one very important stop in San Jose.

I met a man named Wayne Sharpe who is head of a company called Myverse.com and he wanted me to stop and meet him so that he could hire me to work for him. I really love Wayne and this man has dedicated everything to this company. Meeting him was a pleasure I said. All we had to do was drive a hundred miles out of our way. A breeze.

Yeah, until Laura and I stop so she can get clean bandages and some junkies try to rob our car.

The thieves were so brazen that when we confronted them at our car they called the cops on us as a diversion! They had the nerve to report us so the police for abuse so that the police come there to a mall located in Salinas California with the story all wrong.

Once I met the police at the mall and Laura and I told them the real story of what went down, I ended up having a selfie photo with the policeman taken and he gives me his card.

We go and shake that incident off, drive three more hours, meet Wayne and then continue on after having a nice meal with my friend in San Jose. We finally get to Los Angeles and we wonder if things will ever settle the hell down.

The next day in what seemed like only minutes, we had a beach side marriage performed by Anthony with his wife Erica and their sons were there, as well as his brother Daniel and his family attending. They were all there for us when we had no hope and that was going to be how it went. We thought some strangers would help us and no more of it. The girls had nice dresses on and Laura looked perfect. I had at least gotten this gift from our efforts.

That was it then, we got the ceremony over with in a really cool way near the beach, went for a nice meal afterwards and I told Anthony that Laura and I were really grateful then that he was the one who had done all of this for us.

Laura and I had a tow box which we bought new fitted to the car, so that I could load it up with 45 boxes that had 26 books in each one. We then filled it further with all of our camping gear and finally set off for Canada.

I had a cargo box on top of the SUV put on the car filled with the rest of our gear so that we had everything we needed for weeks. Our honeymoon began while we were on our way North along "Pacific Coast Highway 1" and it seemed like we could shake off Los Angeles and the many things there that went badly.

When we said to one another "Let's go back to Pentaluna"!, I should have know it was not going to stop and we had more crazy things coming. It was reinforced near San Luis Obisbo when I saw a motorcycle go straight into the back of truck and I just had enough time to get Laura to avert her eyes.

We awoke in a motel-6 the next day after getting in at 2:am to the hotel because the traffic around San Franciso was jammed. I went outside to get coffee in the office and as I do I saw a pair of dogs in a beat up pick up truck looking anxiously up at a second floor room. I thought nothing of it, just felt badly for the dogs being left outside all night or alone in the truck.

When I went back to the room and handed Laura a coffee for her to drink she saw like 5 police cars all pull up behind me down below.

Laura and I just wanted to get the girls dressed and go to the Santa Rosa courthouse and drop the marriage documents off so we could register it. When we saw the police speaking to a man who obviously owned the truck that had the two dogs in it and we saw a body being wheeled out of a hotel room across from ours, we knew this was bad. A policeman asked me about what I saw and I told him simply that I saw two dogs in the truck he was beside. I said we got there at 2:am and heard and saw nothing. The policeman took my statement and we left with that odd feeling this was bad.

We would read in the news online the next week that a woman had been murdered in that hotel room and the man whom the police were questioning was arrested for it. I don't know if they will contact me further about it all I just really feel badly for the animals and the poor woman who died.

Once again it just kept going from there towards Canada, and more event after event days would occur. I felt like Odysseus in that moment where in his story he decides to go on despite terrible foreboding thoughts, all because love makes me continue.

No Adrian, we are not continuing down this crazy ass mountain road!

The girls in Big Bear before it went badly...

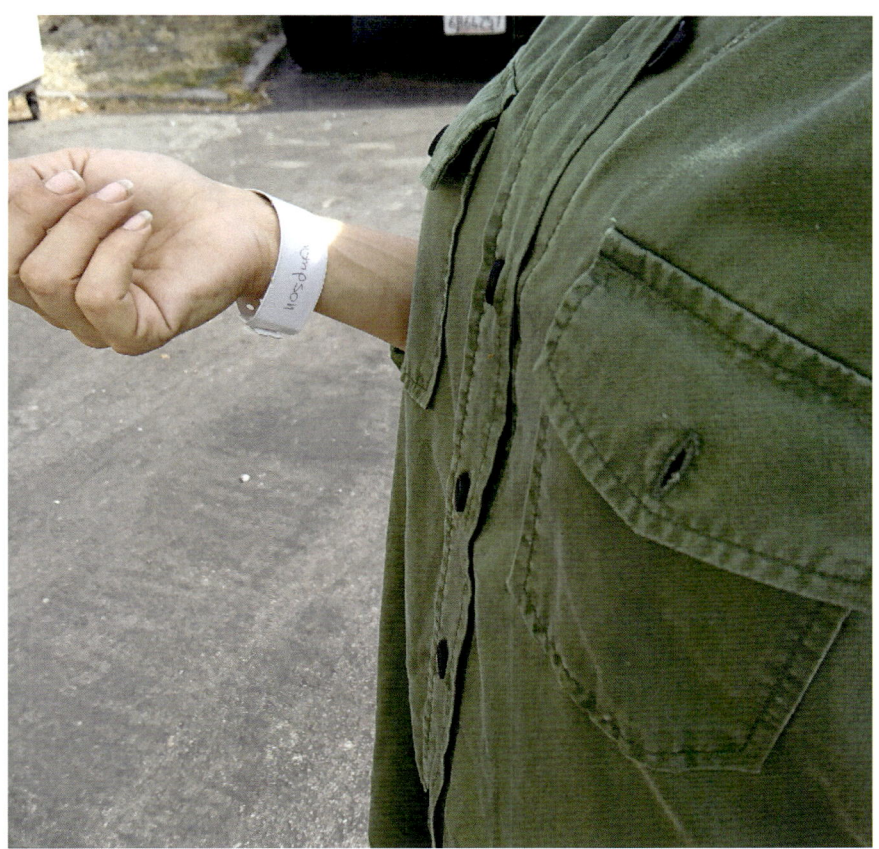

You can reshape yourself just like that.

Getting rescued from a Mountainside by CHP.

All of a sudden Disneyland made more sense than the outside world...

A completed process that took so long to be, but so worth it now.

Love you Anthony, you are a good soul.

The trade off...

My Journey Through *Her* Eyes

We innocently thought this was a cute moment...not a murder scene.
(Blue Truck is one in question.)

We knew how long going back would be...onward we continued north to Canada.

6 FREE WATER AT THE CANADIAN BOARDER!

My goal was to be in Canada one week ahead of when it was that I would be speaking to a company there. It was now late July and I had to be in Calgary by the 13th. Yeah I know me and this number...it does not matter to me, I have all sorts of crazy happen without any numerology involved.

Laura and I left California entered Oregon along the coast. Turn after turn we said "wow"! To the new beauty along the Redwood highway. We alternated between camping and hotels and we had so much time that we got to stop and enjoy ourselves. This part of the trip was a mix of things for us. We learned that a Raccoon can be as big as a dog and they are not afraid to walk across your head in a tent. We saw Elk, Whales, Seals, Eagles, and numerous deer. All around us was green and lush and the heat of the South was long gone. I drove day after day with the girls on their I-pads or singing songs all the way to Washington.

Laura has a sister Kayleigh who has a job that makes her travel to the USA. She was here in America for work and we arranged that when we finished my work in Canada she could meet us in Seattle and we could she could spend a week with her two nieces and sister. I knew how important the girls are to Kayleigh as she fawns over them any chance she gets. I was happy to do whatever it took for this reunion to happen and Laura and I even hoped for a few hours as adults out for dinner while her sister was in town.

One of the things I love about the road is the people you can meet.

We were stopped along a bit of PCH1 Highway on the boarder of Washington when a man just pulled over and asked us if we were going into Washington. When we replied that we were, he handed us a year long pass to all of the Washington State parks. Just like that we had free access to all of the camping we needed.

That one incident showed later why so many things happen for a reason. We were going to only be in Washington State for like 3 days at best as we passed through to Canada. We were going to camp on that side of the border as were crossed the Canadian Rocky mountains.

Seattle is a traffic choked place that is getting worse because it was never meant to be that full of humans. For the sake of not having all manner of troubles we camped in the national campgrounds or did cheap hotels as we made our way to Custer Washington. As we drove up to the Canadian border in the SUV I had this weird feeling I had been there or that it was somehow familiar. When they handed me papers and asked us to pull into the search area, I knew this was going to be a long day.

As I walked up to the counter inside the immigration building a mid-40's blonde dyed haired woman signaled for me to approach her. She asked me what I was doing in Canada and my purpose there.

I tried to explain that I was a professional speaker and that I was there to speak before a company, she just looked like she was not following along. As soon as I started to speak to her she held her hand up over and over as she said that I had to slow down, this was all a bit much for her to digest.

I breathed... then I said as patiently as I could that I was an inspirational speaker, (she quickly interjected; "Well, I never heard of you") so I blinked that one off and continued saying how I was there for a company with 2.500 employees who were expecting me.

You ever meet someone who instantly resumes their domestic dispute mannerisms because you represent a male or a female that that are frequently at odds with? You know exactly what I mean too...that man or woman that you meet who instantly talks to you the same way in which they speak to someone the are constantly arguing with? Yeah well, that was what this woman was acting like. Every time I opened my mouth to answer her she was on me in a domineering way like we were real familiar with one another.

A male Border agent seated next to her saw the exchange and began to sidle into the conversation with placating words. This only seemed to agitate the female agent and I saw right away that he was the one who reigned her in. I was dealing with a queen of the roost type of personality. It was clear when one of her first lines that was delivered in a snide mannerism was: "This is MY country and I don't care if you were here previously, you don't get in unless I say so"!

Wow, all of my efforts to be super polite and expertly mannered in my dealing with her took a serious effort of discipline. I went over and sat with Laura and the girls while this woman looked up my records after she curtly dismissed me so she could "Go have a look what you claim is true".

Laura was confused by all of this as she had seen me get invited or be accepted in very prestigious places.

I never had any issues going to any country. I tried to explain to her that this was a regional thing, that each side of a border punishes neighbors and it would be the same if I was Canadian and going into the USA. Each side is involved in a petty tit-for-tat game when a 70 old grand father from Canada gets his heart broken on his way to Disneyland with grandchildren, all because when he was 17 he was arrested for theft. The "freedom of movement act" of which both countries is party to is a joke. Human beings are not meant to be persecuted for simply tying to visit a country. Any argument about their worthiness is supposed to be engaged once that person declares that they wish to stay some new place.

   Laura was not trying to hear this as she was really hurt to hear this woman start to exert her powers and speak to me in such a callous way. She asked me how can I let people talk to me like this and my heart melted. I began to emotionally break down and I started feeling so humiliated. I went over to the woman and I asked her after 6 hours of being tormented about how I got off of death row and why did I escape and I asked her: Why are you doing this to me miss"? "Why are you making sure to humiliate me at each chance that you get to speak to me"? I asked wasn't I polite enough? Haven't I tried sincerely to answer you?

   As I stood with tears streaming down my face the male officer once again got involved. He asked me to tell him everything and he would do what he could to write a recommendation for me to enter. I went over and sat next to Laura and told her..."Maybe".

   I swear on everything that I value in life that I watched this woman poison my chances of entry. She was so put off that her male colleague and I were kindly interacting that she was doing that facial thing of disdain for his caring mannerisms towards me.

It was like sitting next to your sister on a car journey and you ruined her favorite music with you clumsy ways. All elbows and harsh sounds was how this one went. And like I said, I saw her leap to it while the male border agent was trying to get an approval for my entry form approved. She went off and found my convictions from the NCIC computer records. This is the national criminal arrest records America partially shares with Canada. By partial, I mean just how many times you ever were arrest or convicted. My arrest records sheet prints out like the machine is chewing on a walnut or something. When I saw this female officer with a packet of papers in her hands while motioning for her male counterpart to abandon his 'nicey-nice time' with me and come see what *she* had for him, saying aloud; "I'VE GOT *IT*"!

The male officer excused himself like he was about to go in the back room with her and look at gruesome images of car wrecks with clowns playing with puppy images tossed in to brighten the mood. I sank inside as I went back over to Laura about 8 hours into this ordeal and sat next to her.

I looked at Laura then and I said; "Babe, I really think this woman is so bent on not being "wrong"about me being a lowlife that she is going to do all that she can to stop us". Laura had heard this woman scoff at my UN security clearance, my EU recommendation, my being on the internet in a film. She heard my getting belittled over any and all that I spoke about. She was livid that I got ridiculed or made a joke of.

I was so hurt because this woman knew she was aiming this at Laura in a way women can do, you know, watch me hurt your man while smiling in Laura's face.

About 10 hours into this thing and very late at night, I asked them what about our children going without food or water all this time? We had been made to stay in chairs in a big lobby as we waited out my ordeal about my past history in the control area.

They border agency said that there was no food or water there inside the building and that they had nothing to give us. Our dogs were in crates for hours, we were wrung through emotionally and they were surprised that we even asked about food or water.

An officer was dispatched in a border agency car to go get water and food for the girls. We were given bottles of water from a famous store chain located in Canada that have purity on the label. We were also given ham sandwiches made the day before. I ate nothing and drank nothing.

Keep your water I thought to myself. If you are going to hurt me I will not take alms from your hand while you do your harm. I made sure that the girls drank water and watched as they picked at the food. We sat in silence while we awaited a supervisor to come over from another area of Canada to look over the paperwork which these 2 agents were now preparing for her.

When the female supervisor finally showed up I then went through another trial before her, all about my past and what I was convicted of and what I was innocent of...

By that point I was so emotionally beat down from these people tormenting me about what I was arrested for. I pleaded with them to look at my life since release. I asked what did I possibly do to deserve this? The supervisor told me if I could answer only ONE question, that I could come into Canada.

What happened to my state of Florida convictions? Did I serve all of my Florida time, and what happened to this criminal record eventually because it showed that this file was still "active" some 32 years after it happened.

Laura and I sadly packed up the kids and the dogs and we began to drive 100 yards back into America. As we pulled into their security area on that side we of course had to explain how our Canadian friends just booted us out and why. They happily told us how they picked off "this one" or "that one of theirs", like this was somehow going to make us feel better. I really didn't care that a man who once drove drunk, stole a bike as a teenager, or punched someone in the face over his wife were all stopped from going into America.

We found a hotel in Washington state at about 1:30 am after I had drove us back. I really felt like I had just had my family and I shown what "petty" could feel like. I hate petty. All I could do that night was to e-mail my former Death Row attorney Peter Goldberger back in Pennsylvania and ask ,if by chance, he has my criminal files 14 years after I got out.

I send word to Anthony Samadani who was beside himself over my treatment because he was the one who helped me to arrange this trip. I knew he was taking it hard that my reputation and my speaking career was being hurt by this action.

Peter Goldberger told me that he cannot keep records longer than 7 years but that he has no choice but to hold onto mine because of its significance to him...all 23 boxes he still has at his offices. He said he would help me with Canada by getting me my release papers from Florida and my court papers sent by email to me.

Peter really is the sweetest man, and just like an uncle who wants to be sure that I am alright mentally before he acts, he got my Florida records in hours after he made me tell him every damn thing that went down. By the time I got off the phone with him the next morning he had a lawyer in Canada calling me to see if they could help. Peter is the kind of guy who people leap at a chance to help him in any way they can just out of respect for his lifetime of service to the courts.

   Two days after I went to the Canadian Boarder, I was back there again with Laura and the girls and our dogs. I made sure to get there early, I had water and food as well as the proper documents I needed. I really was so grateful to Peter and all I had to do from there was drive 800 miles across the Rockies in 5 days and I would be selling books and doing my thing speaking.

   My pseudo-ex jilted lover of a female border agent was acting like she did not recognize me as she saw my car being sent to the security area. She busied herself with papers as we got out of the car and walked into her building. She was letting some other agent handle my case and it was all so much of an act she put on to not be her.

   I knew I was waiting to see the female supervisor who told me what I needed to bring back. When I explained what I was told to bring and why I was told that she would come to see me when she began work later on.

   Then began the hours of the children sitting with nothing to do, the people like us sent into the building…some who got arrested right after smiling at our children, they were Handcuffed and taken out of the room and now the girls want to know if they are going to handcuff them and do this to me.

Inside I am just taking blow after blow emotionally and Laura has been worn thin by this all. When the female supervisor finally shows up I literally have to go through each part of my whole story from start to finish and I was not allowed to cut to the "here is your papers you asked for" as it was being slowly recorded by a third member of staff by hand written notes.

   When we got to the criminal records she said that while she had indeed told me to go fetch the paperwork related to my Florida convictions, that now she wanted me to go over each and every time I was arrested since the age 17. My mouth was agape as she said this to me as I really did believe her the other night when she said that all I would need is my discharge papers and the court records showing the Florida convictions were over.

   No, I was given a hand written copy of their notes, I sat for hours and went over with them for 6 more hours of this endless questioning. I was told that when I can come back and have a piece of paper with me for each and every one of the dozen or so queries they now had, that I could then "re-attempt" an entry. I looked at then and I said that they knew and I knew that there was no way I could get all of this done and be back in time to then drive to Calgary and do the job I was hired for. They knew that if I did not enter then that I was losing my speaking job and I was going to be unable to get past the sheer distance of driving in time.

   Canada made me sign a waiver that I was leaving on my own accord. I asked what happened if I refused. They said they would put my wife and children into custody and take the dogs to the animal center for keeping. I thought about the girls seeing me in handcuffs and I knew I could not fight this.

Once again we drove the 100 yards back between countries and into America and had to explain why the country of Canada rejected me. I was so tired of saying the words Death Row that I could burst with the headache I had. I knew I was getting burnt by Canada. I knew I was losing like 50.000.00 dollars or more from this denial of entry. Book sales alone were enormous. The company who agreed to hire me is not paying me any of my expenses that I used to get to Canada. They do not care that it was an issue with my past, they want their money back that I was paid as an advance too.

   What could I do? It is not even like I could go any place because Laura's sister was on her way to Seattle to meet us in a couple of weeks. I could not go drive 2.000 miles each way to go get the RV which we left parked in front of Noah's house in Woodland Hills. What were we going to do about how we now were facing a financial crisis and no place for the girls to be?

   This is that moment. This is what years of abuse or bad events has made me able to handle. Most people break because what lies ahead seems so daunting that they simply quit emotionally and mentally. Me? I only have to look at the faces of those whom I love and I will never quit somehow trying.

   After a few nights of camping with all the critters Washington State has to offer in it's campgrounds, I lay one night looking on "Craigslist" for a place us to settle in and to stop this madness.

   I don't think it, I know that this all became *too* much by then. I had such a love or Southern Oregon that I found a house that was only 16 hours drive from the Canadian border.

I looked at Laura after I showed her pictures of deer in the garden of the new house and how the house was fully furnished. I asked her if she would let me get us this place so that she could heal somehow with the girls. She asked me how far was it away from where we were and I said to her: "only a couple days drive", followed by; "We can be there and have your sister come there and stay with us for a few days".

Next thing you know we are Oregon bound...

Morning coffee with a beautiful deer while my lawyer got us paperwork...

Ever felt like you are caught between rivals? Yeah, me too...

By the third crash ending in fatality, I figured it was time for us to get off of the road.

When Laura saw this from our back steps, she said "yes" to living here. This is now my office view where I wrote this book.

Having known life only on the road these two needed a home, too.

From our Flat in Ilchester we waited and prayed for better many nights.

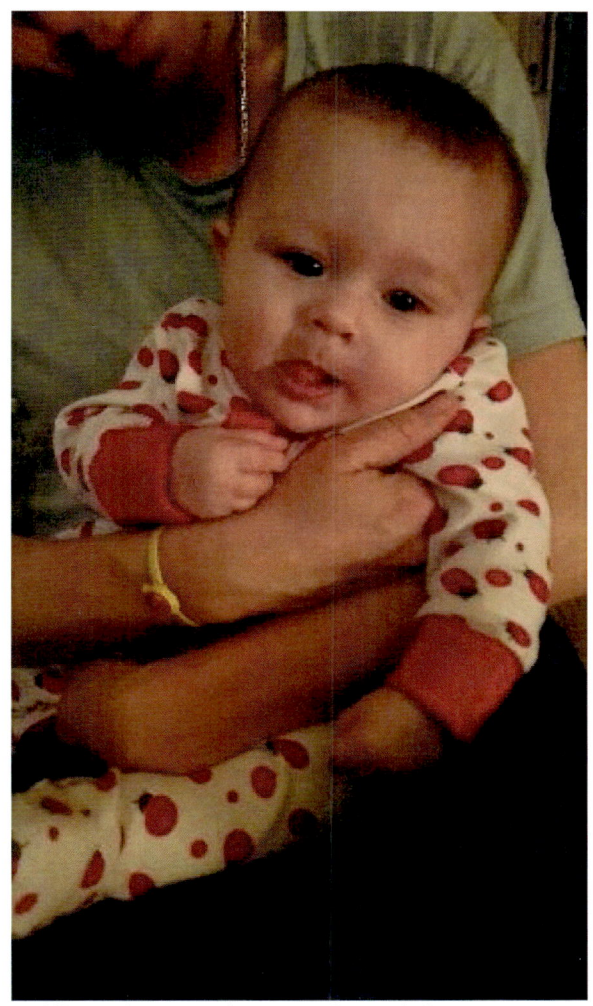

I will continue my journey through *her* eyes now...

One night in Wales I floated on a dream while in love and being happy in these streets. May I always know what that feels like.

7. ALL THE ELEMENTS

Surprisingly I got a response to my efforts on Craigslist to find somewhere that Laura and I could settle in and heal at. The lady on the phone is Alice. She asked us if we wanted to come see the property she had listed for the owners. When we told Alice over the phone that we were in Astoria Oregon on the Washington border, she told us to give her a call when we managed to come down. I don't think she believed we would be on her doorstep the next afternoon. Who drives 16 hours to go see a property, just one? Me. I know it sounds insane but I really had to get us all off of the road. 3 months was enough of this madness for all of us.

Alice knows everyone in the town where we drove to and she met us at the house within minutes of our coming here. We saw on the internet the images, but to be at this property in person it was so full of peace and tranquility. This could be a haven for us, this could be so many things both good and meaningful.

If we wanted to do this thing where we stop running and start living a quiet life, all I had to do was one or two crazy ass long journeys and *then* we could rest. I knew it was going to be more craziness coming our way, because lets' face it for my life it's always "one last time".

Luckily for us Kayleigh could change her flight scheduled for Seattle and she could land in Medford Oregon. All Laura and the girls and I had to do then was drive across Grants Pass to get her at the airport. We settled into the new house and we explored our nearby seaside town and picked out fun places for Kayleigh and the girls to visit while she was here.

We had no idea that a fire caused by lightening was on it's way directly aimed at our home. We watched as smoke filled the skies over us and ask began to fall. As the day of Kayleigh arriving neared we were seriously afraid we now would lose everything to this fire. We watched as section after section got evacuated by the authorities. We were really anxious as the media kept reporting how this was the biggest fire in the USA currently and that is was the top priority. We saw 20.000 acres burn into 150.000 acres in days. That is how many lives were at stake. We could not see any more mountainside or even the skies. We hid inside and said that when her sister came we could drive away from the area possibly.

That was the easily the start of this really anxious time for all of us. The fire cut through the mountains and closed the main road so we had to take a state forestry road.

Now, I do not know how many people have driven on a single lane forestry road before. I can tell you that it is unpaved, sheer drop offs thousands of feet down and you have no choice but to keep up with the car in front or be the one hit by oncoming trucks on the twisty turns or you have to park by the edge and hope you don't go too far as another vehicle squeezes past you.

White knuckled I drove along following another SUV. His plumes of dust nearly obliterated the road I had to balance my distance between losing him or using him as a guide. My gut was tight as I held onto the wheel at 30 miles per hour up and up and more up that mountain.

The side of the road is so sheer Laura was getting nauseous from feelings of vertigo if she looked out of the passenger window.

It was hardest on a turn in the road when another vehicle came along and we did that thing of trying to see who could fit where first so that the other made it by. We had a string of 14 straight trucks come past and each one was towing a trailer from an river expedition that was returning to the coast. I lost my lead car that I was following and the dust he had kicked up had left the car heating up badly and we were nearly out of fuel. No telephone and no maps to help us we drove in a twisting pattern towards a place called Galice Oregon located at the top of National Forestry Road number 23. It was there we would continue on the Grants Pass.

We got all the way through the worst drive ever for children and parents to attempt on the rural road and we are desperate for fuel. I check and I don't have my debit card with me and no cash.

Oh well, everyone here is so nice I was sure that we could be okay. We only needed a gallon of gasoline to make it to Grants Pass where my bank there would let me use my ID to get cash out. We needed like 3.00 dollars of gas.

The only thing is we were in the middle of nowhere and the only place in which we can get fuel is a restaurant/store/gas station there. I politely asked a man servicing boats for the river if he could loan us a gallon of fuel from one of the 30 or so cans saw all over the boat yard. He told me that the really nice lady who ran the place had to say okay to that. He sent me in to talk to her and described her as the boss lady.

Now before I went in, the car is covered in dirt I confess from the crazy ride we just survived, I am obviously not a derelict but I am in jeans and t-shirt. My wife and children are in the car parked right by the front door and I enter the restaurant and approach the counter. I see a woman and ask if she is the manager and when she offers how she is not, I start to relate to her what the service mechanic in the yard outside told me to do about getting a gallon of gas. I was about on the second line when "Boss Lady" cut me off to have me start all over as she twigged on to me speaking to one of her employees who was now shirking away. Oh, she is THAT kind of lady to work for...I get it.

First let me say that I never expected the reaction I got to as politely as I could ask for a simple bit of help. When I told this woman our situation, she shouted in my face that she was calling the sheriff and how I should get the hell out of her place now!

I was stunned. I looked at her and said we got diverted by the fire that is just up the road and we just have to get to a bank for money get gas 20 miles away. I did not ask for a whole tank of fuel from them, just some emergency help for us to get to the airport. This lady came outside after shouting for me to leave and she started telling me how I was not allowed to ask any of her customers for help if that was my plan...I hadn't even tried. Then she shouted at the mechanic in the boat yard not to help us like we were some criminals who had shown up and had stolen from them. I got really tired of her mouth and as she stood on the steps of her crappy restaurant located in backwater that she was queen of, I whipped open the back door of the car and I said: "Here lady, here are my tired and scared children, shout at them that you don't want anyone to help us or you are calling the police because we have a emergency situation...

She said that we were some local scam artist now using our kids for sympathy and Laura held up her British Passport and told the woman she was retarded for thinking we flew half way around the world to rip her off for 3 dollars! The woman only could respond with how we should have had more money with us when we left. She then shouted at her staff to get back inside and someone to please call the police on us to have us thrown out of there.

I looked at Laura and I swear we were stunned that we were not only refused help but we got attacked like that. We even had a 50.00 dollar winning scratch card from Washington state in the car and no one would give us a gallon of fuel for it. Seriously, not one person would help and this woman sure as hell was making sure no one was allowed to help us.

I sure hope her life never holds a moment of needed help and have someone do to her what she did to my children and my wife and I. That was some really mentally disturbed reaction to a family on the road needing help. That whole thing of her owning the only business for miles has made her a really abusive bully and man and woman alike know it. I am damn sure dropping off a copy of this book should I pass by that place again. Hell give some to as many as I can before she gets me booted by the police I guess. My goal is that maybe by giving my books out there she can be shown that underneath it all she is shallow and broken. Someone should also tell her that even an artichoke has a heart boss lady.

As I got into the car following my meeting a very mean woman, it showed now that I still had 50 miles range left. All of that time with the car at a sheer angle made the fuel gauge off. I told Laura we had 48 miles to go Grants Pass and let's ignore the nasty reaction we had just gotten and see if we can make it to a bank.

We did. We just made it to Grants Pass, we got to my bank and I got money out. We got fuel, coffees and food for the kids and made it to Medford Airport only a half hour later than we had hoped. Kayleigh was outside waiting for us while having a cigarette, all as she was brutally ignoring a man next to her as we pulled up to collect her. It was all good then we thought. Part one of the last of the journey making was done. Driving the complete opposite way from Medford we skirted the Galice area as we headed out to the coast of Oregon to where our house is located.

Having the new energy in the mix of her sister being in town really helped to brighten things for Laura. She and her sister have a wit laced bond no one else gets and that is how it should be. Bethany and Zara were going mental with so much auntie love and soon we had three really nice days in a row.

But mother nature was going to mess with us and push us. We had the police come to the door and use a mandatory evacuation order to toss us out. The fire was only a short distance away and they made everyone within our are leave immediately. Our hearts sank as we hurriedly packed the car and connected the tow box with all of our belongings inside. It was already dark and we were two days ahead of our planned departure date to drive to Seattle and drop Kayleigh off at the airport. We hoped that with all of the evacuations that not all of the hotels were booked up.

Town after town all were full in hotel bookings and we ended up in Bandon Oregon and picking a place unlike any hotel I ever stayed at.

The seagull population around this hotel was massive with guano everywhere. Droppings coated the roof, parking lot, and all of the grounds. Even at 11;pm at night you could see seagulls swooping through the lights looking for food from the bins or nearby. They were being fed daily and you can see it.

The Persian gentleman who owned the place filled me in on his history of being the owner of the hotel across the road but that one burned down, on into how he came to America from Iran in 1979 and he had his family here now...by family I found out that he meant the 20 Skunks he was feeding nightly or the dozen or so Raccoon family members feasting there as well.

We were unpacking the car as the gentleman began tossing cat food out to his hordes of what Kayleigh called "Stank Badgers", -which were the skunks- who were dodging right and left, and then there was the "Trash Pandas" as the Raccoon got labeled. We literally were hiding in the car at times while herds of wild animals walked past. Being sprayed by a Skunk is like sticking your head in a clothes tumble dryer at its highest setting and every every nasty piece of underwear ever worn is spinning around with you in it.

With our first night behind us after our stay at the "Ace Ventura Animal Emporium Hotel" as we called it, back I then drove through Oregon and into Washington State.

16 hours later we were in Seattle and at a hotel located only 2 miles from the airport. We had at least made it there one day ahead of when Kayleigh needed to be there for her flight, so I could finally breathe. 3 adults, two children with car seats, and two young puppies is so stressful. I was just so glad that I managed to get Kayleigh to the airport and we only now had to go back home, drop the tow box off if the house did not burn down by then, and proceed to Noah's house to get the RV sold.

I did not even wait a day in Seattle to rest. Soon as Kayleigh was on her flight, we got in the car and drove 500 miles home. We got passes from the Army to enter our road and we dropped the tow box off and got everything cleaned and ready to go 800 miles to Los Angeles.

I don't know how I am able to do this, especially with the pain levels I live with. By this point alone we are still talking about 6 thousand miles I had driven, over 400 hours behind the wheel and so much physical labor of setting up camp, working on RV, or hauling our possessions... I just do know that despite everything I can hold my own even at age 56.

We set course for Bakersfield California to be the half way point for me to stop driving each way. The ride south was easy and we knew lots of it from our drive to Oregon. The girls did well all the way down and as a treat we picked a cheap hotel with a pool just outside of Los Angeles. We did not want to get into LA late at night and get the RV cooled after it sat in over 100 degree heat all day.

The next morning we were all excited to go get "Stormin Norman", which is the nickname we gave the RV.

We were there at 10 before the heat of Woodland Hills becomes too much to be outside. As I pulled up I knew the RV was gone and we were screwed. I was on the phone to the police right off and that is when they told us what happened. It was not stolen and it was not illegally parked, someone had called in a favor and had it taken. What? The officer I spoke to from Topanga Police dept. where the RV was parked told me that a jurisdiction from another area put a tow order on our RV and had it taken away by request of a private party.

What? I really kept saying this over and over too…let me see if I get this straight. I leave my RV parked in front of my friend Noah's house with his permission and someone across the street with a connection in a police force half way across LA can call in a favor and have our RV taken off the street? Yes, the officer said someone had some "juice" to pull that one off and began wondering aloud to himself why they would want to go and do that…oh well, 800 dollars later I can take the RV to Anaheim one last time and ask my friend Lionel if he will sell if for me and then we can just be done.

As I sit here now I have no clue what we will get out of selling the RV and I am putting my faith in a man whom I barely know to help us do that. I do not know how this will play out, I just had to do what I could not to lose everything when the RV got taken from us.

Before I could just get our stuff and go back to Oregon, we had to go find a hotel for Laura and the girls to sit in all day out of the heat.

I planned to get up super early so I could then drive in my own car to Lake Castaic, some 40 miles north of LA. I then was to drop my car there, get in the RV after paying the ransom, only to then drive 60 miles south down to Anaheim where Rod Barr, who is the main screenwriter on the new motion picture could pick me up. I could then drive Rod and I back to my car own car while he sat in the passenger seat with notes he brought to interviewed me for his script writing. We did this in rush hour traffic. I then left Rod at the lake above LA, drove down to Laura and the girls and had tremors in my limbs from dehydration and exhaustion. I had one day left to get the rest done and we could go. We had to get the dogs their shots, Laura to the doctors and we had to get whatever we owned out of the RV and all of it was while on our way out of town.

When we awoke to rain falling on us on our last day it was so appropriate. It never rains in LA in August. We were located to the west of the city and some of the remnants of Hurricane Harvey was being felt as far away as San Bernadino Ca.. I called Anthony Samadani and I said "Every time it rains in Los Angeles someone dies sir". He acknowledged that it is indeed treacherous driving conditions soon as water falls just as an ambulance went by me and was drowning out our words. I told Anthony we were done that day after going to see the doctors for Laura. I said that I would update him when we pulled in to a hotel once we cleared Los Angeles.

From the hotel that morning where we began with us going down to San Juan Capistrano and then back up through Los Angeles to see Noah one last time so that we could eat a meal with my friend.

Then another another lightening storm began just as we got to Lake Castaic, Laura whipped out her Camera and was taking shots of the lightening. Huge ominous looking rain clouds were everyplace and strike after strike of bolts of electricity were hitting the earth. Within minutes we saw a fire crew already on the roadside battling a fire from lightening. All over us the road itself had thick smoke and with the rain falling steady and winds whipping us and the tow box back and forth we prayed that there was no more to this. Just enough already, please. It feels like this just won't end by now.

   We saw that what began as that rain storm coming down on us as we left LA over the top of Los Angeles then became the largest fire in Los Angeles History. It was so symbolic that we were passing through all of the worst things we could have imaged encountering while also having every physical element challenge us as well. There is no way not one part of this wild journey was not meant to be. Whether I understand all of that some day does not matter. It is just one more chapter in a life that has had many like this one.

   The mist came. Sea mist in Oregon is the balance in life I now see. Once night right after we made it back to our place in Oregon we heard all of this dripping outside. The smoke was so thick for days that we could hardly see more than 50 yards. You got peeks at the stars through the smoke, and we also saw no rain hitting the window. So we walked out front and from every place that there was a tree, it was dripping water. The trees were raining down upon the earth, delivering the heavy mist that covered them. It was falling everywhere, a sweet sound that was scented with the smells of the earth coming alive.

The smoke was losing to the mist and we stood together in fascination to the night sounds before us right then. I told Laura the only chance we have now is to take this amazing story and let me turn it into a book and we can hopefully use this to come back from all this devastation or loss.

Laura knows how important my writing is and she pushed me to finish this work so we can try in a way to put this all behind us.

I am here in Oregon looking to go back to speaking and hoping in any way that I can to give these 2 girls a good life. We enrolled Beth in the local school recently. I took her to the bus stop this morning and saw her get on it for the very first time. She turned to me and said: "This is so exciting"!

Before I got in a reply Beth then did machine-gun-style rapid follow ups to me along our driveway to the bus stop about the girls she liked in her new second grade class, or the books she wrote in yesterday with red pens and did I know "this", and did I know "that", followed by how she wants to call her pony Daisy...I did not agree to a pony did I?

Oh well, I don't know how this plays out. I am doing my best to handle what I did to my body for 3 months. I am trying to handle so much actually, because I constantly worry about my ability to care for the girls and Laura. I am worn down but "okay" if you know what I mean?

I don't sleep well because it's still there. I am still feeling that weight in my arms of that sweet child and I am still holding onto that day when everything died there. I just hope that what I am doing is the right thing to do, for surely I know of no other way to be in life.

Thank you anyone who stuck it out and made this far into this story. It really has helped me to appreciate all that I have gone through, and have still. Your buying this book is my only hope to make it in life and for this, I really am indebted to your kindness.

I am not caring of critics, reviews, nor what people think of this work in it's composition. I know that I am limited by my own abilities as a writer, and I find that I am okay with what I produce. Call it a more sincere effort, call it a mess of mistakes, either way it is mine. I wrote all of this work out of a love of storytelling that I have. I proudly own it in both deeds and with words in this way.

There is nothing left for us to do except to now see if we can heal from so many traumas done to us. That takes time *and* being lucky enough to be graced with the mindfulness to enjoy it. I guess that is all any of us have.

I bid you all good days as I complete this work. I am so grateful that I offer this to you as a way to have you appreciate your own struggles or challenges. Ours seem to make Laura and I grow closer and for that, I know it it is all that I can ask for out of my loving family.

Goodbye and much love,

    Nick

Zara and I going to get Beth from the school bus stop atop out road...Bye everyone.

Nick Yarris

ABOUT THE AUTHOR

Nick Yarris is an American author and public speaker who is also the author of "Fear Of 13 Countdown To Execution" (2016) as well as the book "The Kindness Approach" (2017).
He can be found on the web at www.nickyarris.org

Lightning Source UK Ltd.
Milton Keynes UK
UKRC020321071218
333594UK00010B/445